GEO-METRICS II
REVISED 1986 EDITION

LOWELL W. FOSTER
President and Director
Lowell W. Foster Associates, Inc.
Minneapolis, Minnesota

GEO-METRICS II
AS BASED UPON
ANSI Y14.5M–1982 PRACTICES

The Application of Geometric Tolerancing Techniques (using customary inch system)

REVISED 1986 EDITION

ADDISON-WESLEY PUBLISHING COMPANY, INC.

Reading, Massachusetts   Menlo Park, California   New York
Don Mills, Ontario   Wokingham, England   Amsterdam
Bonn   Paris   Milan   Madrid   Sydney   Singapore
Tokyo   Seoul   Taipei   Mexico City   San Juan

Library of Congress Cataloging in Publication Data

Foster, Lowell W.
    Geo-metrics II.

    Includes index.
    1. Drawing-room practice. 2. Tolerance (Engineering)
I. Title.
T352.F68   1985   604.2′4   85-4032
ISBN 0-201-11527-1

This book is based on the author's earlier volumes, *A Treatise on Geometric and Positional Dimensioning and Tolerancing*, originally published in 1963 by Honeywell, Inc.; *A Treatise on Geometric Dimensioning and Tolerancing*, originally published in 1966 by Honeywell, Inc.; *Geometric Dimensioning and Tolerancing—A Working Guide*, published in 1970 by Addison-Wesley Publishing Company, Inc.; *Geo-metrics: The Metric Application of Geometric Tolerancing*, published in 1974 by Addison-Wesley Publishing Company, Inc., and *Geo-metrics II—The Application of Geometric Tolerancing Techniques* (*Using Customary Inch System*) with Addendum, published 1979, revised 1983 by Addison-Wesley Publishing Company, Inc.

The publisher offers discounts on this book when ordered in quantity for special sales. For more information please contact:
    Corporate & Professional Publishing Group
    Addison-Wesley Publishing Company
    Route 128
    Reading, Massachusetts 01867

Cover design by Dick Hannus
Text design revision by ANCO/Boston, Boston, MA
Set in 11 pt. Times Roman by Techna Type, York, PA
Technical art by ANCO/Boston, Boston, MA

9 10–AL–9594939291
Ninth Printing, November 1991

CONTENTS

PREFACE

This text has been prepared to promote worldwide use of geometric dimensioning and tolerancing and to up-grade knowledge on the subject. Its title, *GEO-METRICS II,* attempts to represent and encourage the giant stride being taken by United States industry in its transition toward greater use of national and international standards on this subject. This text also aspires to be of value to all users worldwide as a catalyst for better understanding and to improve the state of the art.

Another major objective of this text is to promote and encourage use of the principles involved as an effective engineering, production, and quality control language or tool which can provide economic and technical advantages.

To further the foregoing objectives, this text is therefore presented in a format suitable for use as a teaching mechanism, both for the classroom and in-industry situation, as well as for providing a daily ready reference for engineering or production work.

Geo-Metrics II is dedicated to the promoting of standardization of engineering drawing techniques in support of the American National Standard "Dimensioning and Tolerancing," ANSI Y14.5M—1982. This text is also dedicated to furthering worldwide efforts through the International Standards Organization and its standards and programs (e.g., ISO/TCIO/SC5 "Dimensioning and Tolerancing)."

The data contained in this book must be considered advisory and are to be used at the discretion of the reader.

The author wishes to express his gratitude to his many colleagues and professional friends in this field of endeavor around the world who also have made contributions to the state of the art. He also wishes to give due credit to the reference documents listed and to those responsible for them.

Finally, the author wishes to express particular gratitude to his wife, Marion, and son, John, for their assistance, patience, and understanding in enduring the often untimely sacrifices of family together time. Without their encouragement and support, this work could not have been completed.

Minneapolis, Minnesota   L.W.F.
June 1985

ACKNOWLEDGMENTS

Credit is gratefully given to the following persons and to the documents used as reference material:

1. American National Standard ANSI Y14.5M–1982. ''Dimensioning and Tolerancing,'' American Society of Mechanical Engineers, New York City, and American National Standards Institute, New York City.

2. American National Standard ANSI Y14.5–1973. ''Dimensioning and Tolerancing for Engineering Drawings,'' American Society of Mechanical Engineers, New York City, and American National Standards Institute, New York City.

3. American National Standard USASI Y14.5–1966. ''Dimensioning and Tolerancing for Engineering Drawings,'' American Society of Mechanical Engineers, New York City, and American National Standards Institute, New York City.

4. International Standards Organization Standards, ISO/R1101—1969, *Tolerances of Form and of Position, Part I, Generalities, Symbols, Indications on Drawings;* ISO/DIS2691, 1972, *Technical Drawings—Tolerances of Form and Position, Part II,* ISO/R1660—1971, *Technical Drawings—Tolerances of Form and Position, Part III, Dimensioning and Tolerances of Profiles.*

5. SAE Aerospace—Automotive Drawing Standards, Sections A 6, 7, and 8, September 1963.

6. Military Standard 8C, 16 October 1963.

7. ''True Position Dimensioning—Rev. 1962,'' Bibeau & Sweet, Scintilla Division, Bendix Aviation Corporation, Sidney, New York.

8. ''Process Standard 9900011,'' Sandia Corporation, Albuquerque, New Mexico.

9. ''Concepts of the True Position Dimensioning System,'' Sandia Corporation, Albuquerque, New Mexico.

10. Ordnance Corps, ''Standard for Dimensioning and Tolerancing ORD 30–1–7.''

11. ''Synopsis of MIL–STD–8B,'' J. V. LaPointe, Honeywell, Inc., Minneapolis, Minnesota.

12. ''Fundamentals of Position Tolerance,'' John V. Liggett, Massey Ferguson, Inc., Detroit, Michigan.

13. Military Standard 8B, 16 November 1959.

14. ''Clarification of Problem Areas of MIL–STD–8B Dimensioning and Tolerancing—code ident 10001 OD15292, Department of the Navy, Bureau of Naval Weapons, 19 June 1961.''

15. ''A Treatise on Geometric and Positional Dimensioning and Tolerancing,'' Lowell W. Foster, Honeywell, Inc., Minneapolis, Minnesota. First printing December 1963; revised edition November 1964.

16. ''A Treatise on Geometric Dimensioning and Tolerancing,'' Lowell W. Foster, Honeywell, Inc., Minneapolis, Minnesota. First printing January 1966; revised edition first printing September 1966, second printing March 1967, third printing July 1967, fourth printing February 1968, fifth printing July 1968.

17. ''Geometric Dimensioning and Tolerancing,'' Lowell W. Foster, Honeywell, Inc., Minneapolis, Minnesota. Society of Automotive Engineers, New York City. SAE Paper 680488.

18. *Geometric Dimensioning and Tolerancing,* Addison-Wesley Publishing Company, A Working Guide, Lowell W. Foster, Minneapolis, Minnesota. Published by Addison-Wesley Publishing Company, 1970.

19. *Geometric Dimensioning and Tolerancing,* A Working Guide with 1973 Addendums, Lowell W. Foster, Minneapolis, Minnesota. Published by Addison-Wesley Publishing Company, 1974.

20. ''True Position Tolerancing—Before and After the Fact,'' Lowell W. Foster, Honeywell, Inc., Minneapolis, Minnesota. American Society of Tool and Manufacturing Engineers, Dearborn, Michigan. ASTME Paper I068-409.

21. *Geo-Metrics: The Metric Application of Geometric Tolerancing,* Lowell W. Foster, Minneapolis, Minnesota. Published by Addison-Wesley Publishing Company, 1974.

22. *Geo-Metrics: The Metric Application of Geometric Tolerancing,* with Addendum as Based upon ANSI Y14.5M–1982 Practices, Lowell W. Foster, Minneapolis, Minnesota. Published by Addison-Wesley Publishing Company, 1982.

23. *Geo-Metrics II,* with Addendum as Based upon ANSI Y14.5M–1982 Practices, Lowell W. Foster, Minneapolis, Minnesota. Published by Addison-Wesley Publishing Company, 1979.

24. *Geo-Metrics II,* with Addendum as Based upon ANSI Y14.5M–1982 Practices, Lowell W. Foster, Minneapolis, Minnesota. Published by Addison-Wesley Publishing Company, 1983, 1979.

25. The Lowell W. Foster Geometric Dimensioning and Tolerancing Module, ''Geo-Calc'' and User's Guide, used with Texas Instrument's Compact Computer (CC40). Foster/Marker/Rennix; Lowell W. Foster Associates, Inc., Minneapolis, Minnesota.

INTRODUCTION

Geometric dimensioning and tolerancing is a means of specifying engineering design and drawing requirements with respect to actual "function" and "relationship" of part features. Furthermore, it is a technique which, properly applied, ensures the most economical and effective production of these features. Thus geometric dimensioning and tolerancing can be considered both an engineering design drawing language and a functional production and inspection technique. Uniform understanding and interpretation among design, production, and inspection groups are the major objectives of the system. This text discusses the subject step by step, focusing on practical application. Before presenting these details, however, we wish to provide the reader with a brief overview of the basis and status of geometric dimensioning and tolerancing.

The authoritative document governing the use of geometric dimensioning and tolerancing in the United States is ANSI Y14.5M–1982, "Dimensioning and Tolerancing." This standard evolved out of a consolidation of earlier standards, ANSI Y14.5–1973, USASI Y14.5–1966, ASA Y14.5–1957, SAE Automotive Aerospace Drawing Standards (Sections A6, 7 and 8– September 1963) and MIL–STD–8C, October 1963. This consolidation was accomplished over many years by committee action representing military, industrial, and educational interests. The work of the committee has had and continues to have three prime objectives:

1. to provide a single standard for practices in the United States,

2. to update existing practices in keeping with technological advances and extend the principles into new areas of application,

3. to establish a single basis and "voice" for the United States in the interest of international trade, in keeping with the United States' desire to be more active, gain greater influence, and pursue a more extensive exchange of ideas with other nations in the area of international standards development.

The historical evolution of geometric dimensioning and tolerancing in the United States is an interesting story which, however, is not discussed in this text. It suffices to say that the early introduction of functional gaging, giving rise to the possibility of new techniques, along with the growing need for more specifically and economically stated engineering design requirements, has caused its growth. Advancing product sophistication and complexity, rapid industrial expansion, diversification, etc., have created an environment in which more exacting engineering drawing communication is not only desirable but mandatory for competitive and effective operation.

Updated and expanded practices have been initiated in the present Y14.5 standard. Further expansion will no doubt occur as growth in the area continues. In the process of extending into new areas, this expansion is confronted by the challenge of ensuring progress without upsetting stability. Rapid advances in this subject, although desirable, must be tempered by the ability to make the transition with no loss of continuity or understanding.

United States coordination and compatibility with international dimensioning and tolerancing practices have been extended significantly in the current Y14.5M–1982 standard.

The symbology and influence from ISO (International Standards Organization) and ABCA (America, Britain, Canada, Australia) documents, activities, and committees have found their way into the current Y14.5 standard. This influence continues in the ongoing development and use of geometric tolerancing in the United States.

Many of the United States industrial concerns and the military have overseas affiliations. Thus, the increasing need for understanding and for more uniform practices throughout the world is evident.

This text presents the subject of dimensioning and tolerancing in order of complexity of the details, and attempts to clarify and promote the use of Y14.5. It also emphasizes the importance of the ongoing effort to expand the principles and to more closely relate to international practices.

GLOSSARY

Actual Size—An actual size is the measured size of the feature.

Angularity—Angularity is the condition of a surface, axis, or center plane which is at a specified angle (other than 90°) from a datum plane or axis. Symbol: ∠.

Attitude Tolerance—See Orientation Tolerance.

Basic Dimension—A dimension specified on a drawing as BASIC (or abbreviated BSC) is a theoretically exact value used to describe the exact size, profile, orientation, or location of a feature or datum target. It is used as the basis from which permissible variations are established by tolerances in feature control frames or on other dimensions or notes. A basic dimension is symbolized by boxing it: 1.270 .

Basic Size—The basic size is that size from which limits of size are derived by the application of allowances and tolerances.

Bilateral Tolerancing—A bilateral tolerance is a tolerance in which variation is permitted in both directions from the specified dimension, e.g., 1.500 ± .005.

Center Plane—Center plane is the middle or median plane of a feature.

Circular Runout—Circular runout is the composite control of circular elements of a surface independently at any circular measuring position as the part is rotated through 360°. Symbol: ⟋.

Circularity (See also Roundness)—Circularity is the condition on a surface of revolution (cylinder, cone, sphere) where all points of the surface intersected by any plane (1) perpendicular to a common axis (cylinder, cone) or (2) passing through a common center (sphere) are equidistant from the center. Symbol: ○.

Clearance Fit—A clearance fit is one having limits of size so prescribed that a clearance always results when mating parts are assembled.

Coaxiality—Coaxiality of features exists when two or more features have coincident axes, i.e., a feature axis and a datum feature axis.

Concentricity—Concentricity is a condition in which two or more features (cylinders, cones, spheres, hexagons, etc.) in any combination have a common axis. Symbol: ◎ (or, formerly, ●).

Contour Tolerancing—See Profile of a Line or Surface.

Cylindricity—Cylindricity is a condition of a surface of revolution in which all points of the surface are equidistant from a common axis. Symbol: $\lozenge$.

Datum—A theoretically exact point, axis, or plane derived from the true geometric counterpart of a specified datum feature. A datum is the origin from which the location or geometric characteristics of features of a part are established.

Datum Axis—The datum axis is the theoretically exact axis of the datum feature (a center line on the drawing) and the axis of the actual datum feature when its surface is in contact with the simulated datum; the smallest circumscribed cylinder (for external features) or largest inscribed cylinder (for internal features).

Datum Feature—A datum feature is an actual (physical) feature of a part used to establish a datum.

Datum Feature Symbol—The datum feature symbol contains the datum reference letter in a drawn rectangular box, e.g., $\boxed{-A-}$.

Datum Line—A datum line is that which has length but no breadth or depth such as the intersection line of two planes, center line or axis of holes or cylinders, reference line for tooling, gaging, or datum target purposes.

Datum Plane—A datum plane is a theoretically exact plane established by the extremities or contacting points of the actual feature surface with a reference plane (surface plate or other checking device, i.e., the simulated datum) or a plane established by datum targets.

Datum Point—A datum point is that which has position but no extent such as the apex of a pyramid or cone, center point of a sphere, or reference point on a surface for tooling, gaging, or datum target purposes.

Datum Reference—A datum reference is a datum feature and the resulting datum plane or axis.

Datum Reference Frame—A datum reference frame is a set of three mutually perpendicular datum planes or axes established from the simulated datums in contact with datum surfaces or features and used as a basis for dimensions for design, manufacture, and measurement. It provides complete orientation for the features involved.

Datum Surface—A datum surface or feature (hole, slot, diameter, etc.) refers to the actual part, surface, or feature coincidental with, relative to, and/or used to establish a datum plane.

Datum Target—A datum target is a specified point, line, or area (identified on the drawing with a datum target symbol) used to establish datum points, lines, planes, or areas for special function, or manufacturing and inspection repeatability. Symbol: (A1)

Dimension—A dimension is a numerical value expressed in appropriate units of measure and indicated on a drawing and in other documents along with lines, symbols and notes to define the size or geometric characteristic (or both) of a part or part feature.

Feature—A feature is the general term applied to a physical portion of a part and may include one or more surfaces such as holes, pins, screw threads, profiles, faces, or slots. A feature may be ''individual'' or ''related.''

Feature Control Frame—The feature control frame is a rectangular box containing the geometric characteristic symbol and the form, orientation, profile, runout, or location tolerance. If necessary, datum references and modifiers applicable to the feature or the datums are also contained in the box, e.g. ⟨ ⟩ .002 A .

Feature of Size—A feature of size is one cylindrical or spherical surface, or a set of two plane parallel surfaces, each of which is associated with a size dimension.

Fit—Fit is the general term used to signify the range of tightness or looseness which may result from the application of a specific combination of allowances and tolerance in the design of mating part features. Fits are of four general types: clearance, interference, transition, and line.

Flatness—Flatness is the condition of a surface having all elements in one plane.
Symbol: ⬜.

Form Tolerance—A form tolerance states how far an actual surface or feature is permitted to vary from the desired form implied by the drawing. Expressions of these tolerances refer to flatness, straightness, circularity, and cylindricity, and occasionally profile of surface and profile of line.

Full Indicator Movement (FIM) (FIR) (TIR)—Full indicator movement is the total movement of an indicator (or comparable measuring device) when appropriately applied to a surface to measure its variation.

Full Indicator Reading (FIR) (TIR) (FIM)—See Full Indicator Movement.

Geometric Characteristics—Geometric characteristics refer to the basic elements or building blocks which form the language of geometric dimensioning and tolerancing. Generally, the term refers to all the symbols used in form, orientation, profile, runout, and location tolerancing.

Geometric Tolerance—Geometric tolerance is the general term applied to the category of tolerances used to control form, profile, orientation, runout and location.

Geometrics—Geometrics is a colloquial term to describe geometric tolerance (above).

Implied Datum—An implied datum is an unspecified datum whose influence on the application is implied by the dimensional arrangement on the drawing, e.g., the primary dimensions are tied to an edge surface; this edge is implied as a datum surface and plane.

Interference Fit—An interference fit is one having limits of size so prescribed that an interference always results when mating parts are assembled.

Least Material Condition (LMC)—Least material condition is the condition in which a feature of size contains the least amount of material within the stated limits of size: for example, maximum hole diameter and minimum shaft diameter. It is opposite to Maximum Material Condition (MMC). Symbol: Ⓛ.

Limit Dimensions (Tolerancing)—In limit dimensioning only the maximum and minimum dimensions are specified. When used with dimension lines, the maximum value is placed above the minimum value, e.g., $\frac{.300}{.295}$. When used with leader or note on a single line, the minimum limit is placed first, e.g., .295–.300.

Limits of Size—The limits of size are the specified maximum and minimum sizes of a feature.

Line Fit—A line fit is one having limits of size so prescribed that surface contact or clearance may result when mating parts are assembled.

Location Tolerance—A location tolerance states how far an actual feature may vary from the perfect location implied by the drawing as related to datums or other features. Expressions of these tolerances refer to the category of geometric characteristics containing position, concentricity, and position as used on symmetrical parts.

Maximum Dimension—A maximum dimension represents the acceptable upper limit. The lower limit may be considered any value less than the maximum specified.

Maximum Material Condition (MMC)—Maximum material condition is the condition in which a feature of size contains the maximum amount of material within the stated limits of size: for example, minimum hole diameter and maximum shaft diameter. Symbol: Ⓜ.

Median Plane—Median plane is the middle or center plane of a feature.

Minimum Dimension—A minimum dimension represents the acceptable lower limit. The upper limit may be considered any value greater than the minimum specified.

Minimum Material Condition—See Least Material Condition.

Modifier—A modifier is the term sometimes used to describe the application of the ''maximum material condition,'' ''regardless of feature size,'' or ''least material condition'' principles. The modifiers are maximum material condition (MMC), symbol Ⓜ, regardless of feature size (RFS), symbol Ⓢ, and least material condition (LMC), symbol Ⓛ.

Multiple Datum Reference Frame—A multiple datum reference frame is the condition in which there are one or more related or separate datum frames.

Nominal Size—The nominal size is the stated designation which is used for the purpose of general identification, e.g., 1.400, .060, etc.

Normality—See Perpendicularity.

Orientation Tolerance—An orientation tolerance controls the orientation of features to one another. Expression of these tolerances refers to perpendicularity, angularity, and parallelism, and in some instances, profile, as related to datum features. Orientation tolerances are sometimes referred to as attitude tolerances.

Parallelepiped—Shape of tolerance zone. The term is used where TOTAL WIDTH is required and to describe geometrically a square or rectangular prism, or a solid with six faces, each of which is a parallelogram.

Parallelism—Parallelism is the condition of a surface, line, or axis which is equidistant at all points from a datum plane or axis. Symbol: //.

Perpendicularity—Perpendicularity is the condition of a surface, axis, or line which is 90° from a datum plane or a datum axis. Symbol: ⊥.

Position Tolerance—A position tolerance (formerly called true position tolerance) defines a zone within which the axis or center plane of a feature is permitted to vary from true (theoretically exact) position. Symbol: ⊕.

Profile of a Line—Profile of a line is the condition permitting a uniform amount of profile variation, either unilaterally or bilaterally, along a *line* element of a feature. Symbol: ⌒.

Profile of a Surface—Profile of a surface is the condition permitting a uniform amount of profile variation, either unilaterally or bilaterally, on a *surface*. Symbol: ⌒.

Projected Tolerance Zone—A projected tolerance zone is a tolerance zone applied to a hole in which a pin, stud, screw, or bolt, etc., is to be inserted. It controls the perpendicularity of the hole to the extent of the projection from the hole and as it relates to the mating part clearance. The projected tolerance zone extends *above* the surface of the part to the functional length of the pin, screw, etc., relative to its assembly with the mating part. Symbol: Ⓟ.

Regardless of Feature Size (RFS)—Regardless of feature size is the term used to indicate that a geometric tolerance or datum reference applies at any increment of size of the feature within its size tolerance. Symbol: Ⓢ.

Roundness (See also Circularity)—Roundness is the condition on a surface of revolution (cylinder, cone, sphere) where all points of the surface intersected by any plane (1) perpendicular to a common axis (cylinder, cone) or (2) passing through a common center (sphere) are equidistant from the center. Symbol: ○.

Runout—Runout is the composite deviation from the desired form of a part surface of revolution during full rotation (360°) of the part on a datum axis. Symbols: ↗ or ↗↗.

Runout Tolerance—Runout tolerance states how far an actual surface or feature is permitted to deviate from the desired form implied by the drawing during full rotation of the part on a datum axis. There are two types of runout: circular runout and total runout.

Simulated Datum—A simulated datum is found in the associated processing equipment used to establish a datum axis or plane. For example, machine tables, surface plates, collets, chucks, etc., though not true cylinders or planes, are of such quality that they are used to simulate the datum (planes, axes) from which measurements are taken and dimensions verified when brought into contact with a datum feature.

Size Feature—A size feature is a three dimensional feature, such as a hole or pin (*not* a single plane surface).

Size Tolerance—A size tolerance states how far individual features may vary from the desired size. Size tolerances are specified with either unilateral, bilateral, or limit tolerancing methods.

Specified Datum—A specified datum is a surface or feature identified with a datum feature symbol or note.

Squareness—See Perpendicularity.

Straightness—Straightness is a condition where an element of a surface or an axis is a straight line. Symbol: —.

Symmetry—Symmetry is a condition in which a feature (or features) is symmetrically disposed about the center plane of a datum feature. Symbol: ⌖; or, formerly, ≡.

Tolerance—A tolerance is the total amount by which a specific dimension may vary; thus, the tolerance is the difference between the maximum and minimum limits.

Total Indicator Reading (TIR) (FIR) (FIM)—See FIM, Full Indicator Movement.

Total Runout—Total runout is the simultaneous composite control of all elements of a surface at all circular and profile measuring positions as the part is rotated through 360°. Symbol: ⟋⟋.

Transition Fit—A transition fit is one having limits of size so prescribed that either a clearance or an interference may result when mating parts are assembled.

True Position—True position is the theoretically exact location of a point, line, or plane of a feature in relationship with a datum reference or other feature.

Unilateral Tolerance—A unilateral tolerance is a tolerance in which variation is permitted only in one direction from the specified dimension, e.g., $1.400 \; {}^{+.000}_{-.005}$.

Virtual Condition (Size)—Virtual condition (size) of a feature is the boundary generated by the collective effects of the specified MMC limit of size of a feature and any applicable geometric tolerances.

WHY USE GEOMETRIC DIMENSIONING AND TOLERANCING?

Why is it that we should be so interested in this subject?

FIRST AND FOREMOST ITS USE SAVES MONEY!

It saves money directly by providing for maximum producibility of the part through maximum production tolerances. It provides ''bonus'' or extra tolerances in many cases.

It ensures that design dimensional and tolerance requirements, as they relate to actual function, are specifically stated and thus carried out.

It adapts to, and assists, computerization techniques in design and manufacture.

It ensures interchangeability of mating parts at assembly.

It provides uniformity and convenience in drawing delineation and interpretation, thereby reducing controversy and guesswork.

Aside from these primary reasons there are others of a more general nature:

The intricacies of today's sophisticated engineering design demand new and better ways of accurately and reliably communicating requirements. Old methods simply no longer suffice.

Diversity of product line and manufacture makes considerably more stringent demands of the completeness, uniformity, and clarity of drawings.

It is increasingly becoming the ''spoken word'' throughout industry, the military, and, internationally, on engineering drawing documentation. Every engineer or technician involved in originating or reading a drawing should have a working knowledge of this new state of the art.

WHAT IS GEOMETRIC DIMENSIONING AND TOLERANCING?

In particular, it is a means of dimensioning and tolerancing a drawing with respect to the actual function or relationship of part features which can be most economically produced. *Function* and *relationship* are the key words.

In general, it is a system of building blocks for good drawing practice which provides the means of stating necessary dimensional or tolerance requirements on the drawing not otherwise covered by implication or standard interpretation.

WHEN SHOULD GEOMETRIC DIMENSIONING AND TOLERANCING BE USED?

When part features are critical to function or interchangeability;

when functional gaging techniques are desirable;

when datum references are desirable to ensure consistency between manufacturing and gaging operations;

when computerization techniques in design and manufacture are desirable;

when standard interpretation or tolerance is not already implied.

GEOMETRIC CHARACTERISTICS AND SYMBOLS

The geometric characteristics and symbols that are used as the building blocks for geometric dimensioning and tolerancing are:

Symbol	Characteristic
▱	Flatness
—	Straightness
○	Circularity (Roundness)
⌀	Cylindricity
⊥	Perpendicularity (Squareness)
∠	Angularity
//	Parallelism
⌒	Profile of a Surface
⌒	Profile of a Line
↗	Circular Runout
↗↗	Total Runout
⊕	Position
◎	Concentricity
⹀	Symmetry (replaced by Position in 1982 standard)

USING SYMBOLS

The general use of symbols instead of notes on a drawing provides a number of advantages. The illustrations below incorporate the geometric characteristic symbols with datum and feature control symbols. Some of the advantages of symbols over notes are

1. The symbol has uniform meaning. A note can be stated inconsistently, with a possibility of misunderstanding.

2. Symbols are compact, quickly drawn, and can be placed on the drawing where the control applies; symbols adapt readily to computer applications.

   Notes require much more time and space, tend to be scattered on the drawing, often appear as footnotes which separate the note from the feature to which it applies.

3. Symbols are the international language and surmount individual language barriers.

   Notes may require translation if the drawing is used in another country.

4. Symbols can be applied with drafting templates or computer techniques and retain better legibility in various forms of copy reproduction.

5. Geometric tolerancing symbols follow the established precedent of other well-known symbol systems, e.g., electrical and electronic, welding, surface texture, etc.

USING SYMBOLS

USING NOTES

△2 SURFACE C PERPENDICULAR TO DATUM A WITHIN .001 & WITH DATUM B WITHIN .001

△1 SURFACE B PERPENDICULAR TO DATUM A WITHIN .001

9

MAXIMUM MATERIAL CONDITION PRINCIPLE
SYMBOL Ⓜ ABBREVIATION (MMC)

One of the fundamental and most important principles of geometric dimensioning and tolerancing is MAXIMUM MATERIAL CONDITION. A thorough understanding of its meaning is therefore essential.

Note in the figure below that the "maximum material condition" size of the .250 ± .005 diameter hole is .245, or its *low* limit size. The hole at its low limit obviously retains more material than if it were at its *high* limit or larger size; thus the term "maximum material condition" defines the *low* limit when it applies to a hole or similar feature.

Note similarly that the .235 diameter pin is at its "maximum material condition" size when it is at its *high* limit of size of .240. In this instance it is more readily seen that more material exists in the pin when it is at its maximum permissible size. However, the same principle exists in both hole and pin MMC situations. Relating mating part features in this manner ensures their functional relationships, and as will be seen later in the text, establishes the criteria for determining necessary form, orientation, and position tolerances.

The symbol for "maximum material condition," the M enclosed in a circle, and the occasionally used abbreviation MMC are shown above. The symbolic method is to be used with feature control frames only. The abbreviation MMC may be used with note callouts but not with symbolic representations. We shall discuss later the application of the "maximum material condition" principle and illustrate it with practical examples.

Generally, the use of the "maximum material condition" principle permits greater possible tolerance as part feature sizes vary from their calculated "maximum material condition" limits. It also ensures interchangeability and permits functional gaging techniques. It is one of the fundamental principles upon which the system of geometric dimensioning and tolerancing is based. Below is the definition of maximum material condition and the usual prerequisites for application. We shall later expand the use of the principle by means of examples.

Definition. The condition in which a feature of size contains the maximum amount of material within the stated limits of size: for example, minimum hole diameter and maximum shaft diameter.

The "maximum material condition" principle is normally valid only when both of the following conditions exist:

1. Two or more features are interrelated with respect to location or orientation (e.g., a hole and an edge or surface, two holes, etc.). At least *one* of the related features is to be a feature of size.

2. The feature (or features) to which the MMC principle is to apply must be a feature of size (e.g., a hole, slot, pin, etc.) with an axis or center plane.

"Maximum material condition" might also be considered as a "new" term for an "old" situation, such as the familiar terms "worst condition," "critical size," etc., used in the past for relating mating part features.

Where the maximum material condition principle is not appropriate, the "regardless of feature size" principle may be applied. (See below.)

REGARDLESS OF FEATURE SIZE
SYMBOL Ⓢ ABBREVIATION (RFS)

Definition. The term used to indicate that a geometric tolerance or datum reference applies at any increment of size of the feature within its size tolerance.

"Regardless of feature size" is another principle of geometric dimensioning and tolerancing which must be well understood. Unlike maximum material condition, the "regardless of feature size" principle permits *no* additional positional, form or orientation tolerance, no matter to which size the related features are produced. It is really the independent form of dimensioning and tolerancing which has always been used prior to the introduction of the MMC principle.

The symbol for "regardless of feature size" and the occasionally used abbreviation RFS is shown above. We shall later clarify the principle by means of examples.

The RFS principle is valid only when applied to features of size (for example, a hole, slot, pin, etc., with an axis or center plane). The size connotation cannot be applied to a feature which does not have "size."

LEAST MATERIAL CONDITION PRINCIPLE
SYMBOL Ⓛ ABBREVIATION (LMC)

Definition. The condition in which a feature of size contains the least amount of material within the stated limits of size: for example, maximum hole diameter and minimum shaft diameter.

The least material condition principle may be desirable as an alternative to MMC or RFS in certain design considerations. See "Least Material Condition," page 144.

BASIC AND DATUM

The terms BASIC and DATUM are most important. Proper application of the principles implied by these terms greatly contributes to effective geometric dimensioning and tolerancing.

BASIC

Definition. A numerical value used to describe the theoretically exact size, profile, orientation, or location of a feature or datum target. It is used as the basis from which permissible variations are established by tolerances on other dimensions or in notes.

Use of a BASIC dimension, which is a theoretical and exact value, requires also a *tolerance* stating the permissible variation from this exact value (most often relative to a position angularity or profile requirement). A BASIC dimension states only *half* the requirement. To complete it, a tolerance must be associated with the features involved in the BASIC dimension.

A BASIC dimension may be identified on the drawing by the word BASIC (or the accepted abbreviation BSC) adjacent to, or below, the dimension, or by a general note on the drawing. The symbolic method that follows on page 13 is the recommended method in keeping with latest standards.

EXAMPLE

Some companies use a "naked" or untoleranced dimension instead of BASIC. The same meaning as BASIC is invoked by adding a drawing footnote, title block notation, or company standards.

The term TP (for True Position) derived from British standards has also been used in the past. It has the same meaning as BASIC.

BASIC dimensions are also used in applications such as tapers for which tolerances must be derived from associated size dimensions.

The use of BASIC dimensions on datum targets assumes standard tooling or gagemaker's tolerances (see DATUM section for more detail).

SYMBOLIC METHOD OF STATING A BASIC OR THEORETICAL EXACT VALUE—RECOMMENDED

The preferred method of stating an "exact" value replacing BASIC, BSC, TP, etc., is the international (ISO) method recommended by ANSI Y14.5. According to this method, the exact value is enclosed in a frame or box (see example below). The meaning is the same as that of BASIC.

EXAMPLE

The symbolic method for "exact" values may be used with symbolic *or* notated geometric tolerancing. Because of the need to standardize U. S. practices and encourage compatibility with international practices, using the symbolic or boxed "exact" value is strongly recommended.

DATUMS AND DATUM FEATURE SYMBOL

Definitions. A theoretically exact point, axis, or plane derived from the true geometric counterpart of a specified datum feature. A datum is the origin from which the location or geometric characteristics of features of a part are established.

Datum surfaces and datum features are actual part surfaces or features used to establish datums. They include all the surface or feature inaccuracies.

To identify a feature as a datum, we use the following datum feature symbol:

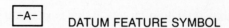   DATUM FEATURE SYMBOL

Each datum requiring identification is assigned a different reference letter. Do not use letters I, O, Q. If the single letter alphabet is exhausted, double letters may be used, i.e., AA, AB, etc.

Note the following conventions:

1.   If the datum feature symbol is shown on the *extension line* of a feature, the datum is for that feature only (see Figs. 1 and 2).

2.   If the datum feature symbol is shown on a *dimension line*—below or adjacent to a DIMENSION, with a leader, note or feature control frame—it applies to the entire dimension or feature with which it is associated (see Fig. 3).

PLACEMENT OF THE DATUM FEATURE SYMBOL

APPLICATION TO PLANE SURFACES (OR CYLINDERS)

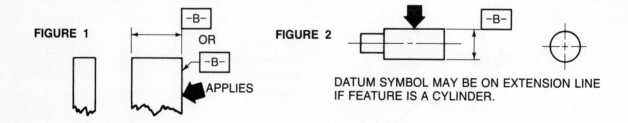

FIGURE 1   APPLIES

FIGURE 2   DATUM SYMBOL MAY BE ON EXTENSION LINE IF FEATURE IS A CYLINDER.

APPLICATION TO SIZE FEATURES

FIGURE 2

APPLIES

FIGURE 3

FEATURE AND FEATURE CONTROL FRAME

FEATURE

The general term applied to a physical portion of a part, such as a surface, hole, or slot.

Features are specific component portions of a part and may include one or more surfaces such as holes, faces, screw threads, profiles, or slots. Features may be ''individual'' or ''related.''

FEATURE CONTROL FRAME

The feature control frame consists of a box containing the geometric characteristic symbol, datum references, tolerance, and the material condition symbol (e.g., for MMC) symbol if applicable.

The following illustration shows a typical feature control frame.

THIS FEATURE:

The example below shows this feature control frame as used on a part drawing.

PLACEMENT OF THE FEATURE CONTROL FRAME

The feature control frame is associated with the feature(s) being toleranced by one of the following methods:

1. Attaching a side, end, or corner of the symbol box to an EXTENSION LINE or leader from the feature (used on most form tolerances). See Fig. 1.

2. Attaching a side or end of the symbol box to the DIMENSION LINE or EXTENSION LINE pertaining to the feature when it is cylindrical. See Fig. 2.

3. Placing the symbol box below or closely adjacent to the dimension or note pertaining to that feature. See Fig. 3.

4. Running a leader line from the symbol to the feature. See Fig. 4.

FIGURE 1

FIGURE 2

FIGURE 3   **FIGURE 4**

COMBINED FEATURE CONTROL FRAME AND DATUM FEATURE SYMBOL

When a feature serves as a datum and is also controlled by a geometric tolerance, the feature control frame and the datum feature symbol may be combined as shown.

REFERENCE TO DATUM

When an orientation, profile, runout, or location tolerance must be related to a datum, this relationship is stated by placing the datum reference letter following the geometric characteristic symbol and the tolerance.

The illustrations on page 19 show additional examples of the feature control frames with reference to datums.

Figure 1 is a typical feature control frame using a single datum reference. The symbol reads "This feature shall be within a .002 tolerance zone perpendicular to datum A."

Figure 2 shows a feature control frame with *two* datums. The symbol reads "This feature shall be located at true position within .005 diameter at maximum material condition with respect to both datums A and B."

Note that vertical lines are used to separate the characteristic symbol, the feature tolerance, and the datum references. These vertical lines are used on all feature control frames to ensure clarity. One reason for this is illustrated in Fig. 3, in which the maximum material condition symbol is used. The vertical lines clearly show that the MMC condition symbols apply only to those datums or tolerances with which they appear in the subdivision of the symbol box.

Figure 4 illustrates primary, secondary, and tertiary datums showing the order of precedence. When the order of precedence of datums is significant to function, datum references should be classified as primary, secondary, and tertiary. The datum precedence is shown by placing each datum reference letter in the proper order. The first datum letter (left to right) is considered the primary datum, the second letter secondary, and the third letter tertiary. Thus the datum reference letters will not necessarily be in alphabetical order. See section on DATUMS for further explanation.

Figure 5 illustrates a feature control frame in which multiple datum features are used simultaneously to establish a single datum reference (equal precedence of datum features) e.g., to establish a common datum axis. See section on DATUMS and RUNOUT for further details.

Figure 6 illustrates a possibly questionable use of a datum reference. Note that datum A applies at MMC, whereas the feature controlled applies at RFS. This means that the datum reference is subject to variation and cannot serve as a fixed reference for any RFS relationship. Although there may be exceptions under special circumstances, generally, wherever MMC is used on any datum, the feature controlled should also be controlled at MMC.

DIAMETER SYMBOL ⌀

The symbol used to designate a diameter (DIA) or cylindrical feature or tolerance zone is shown above. The symbol (⌀) precedes the specified tolerance. See Figs. 2, 3, and 4 at right. The symbol is used elsewhere on the drawing to indicate that the feature is circular or cylindrical. The symbol (⌀) precedes the stated nominal size value of the concerned feature. See figures on page 23 and throughout this text.

FIGURE 1

SYMBOL
TOLERANCE
DATUM REFERENCE

FIGURE 2

TWO REFERENCE DATUMS

FIGURE 3

MMC APPLICABLE TO DATUM
MMC APPLICABLE TO TOLERANCE
DATUM REFERENCE

FIGURE 4   ORDER OF PRECEDENCE OF DATUMS

PRIMARY DATUM REFERENCE
SECONDARY DATUM REFERENCE
TERTIARY DATUM REFERENCE

FIGURE 5

MULTIPLE DATUM FEATURES USED SIMULTANEOUSLY TO ESTABLISH A SINGLE DATUM REFERENCE, FOR EXAMPLE, TO ESTABLISH DATUM AXIS.

FIGURE 6

FEATURE CONTROL RFS
DATUM MMC

NOT NORMALLY PRACTICAL

GEOMETRIC CHARACTERISTICS—
FORM, ORIENTATION,
PROFILE, RUNOUT, AND LOCATIONAL
TOLERANCE—OTHER SYMBOLS AND TERMS

Illustrated at right are the geometric characteristics and symbols which are the basis for the language of geometric dimensioning and tolerancing.

FIVE TYPES OF GEOMETRIC CHARACTERISTICS

Expanding on preceding text explanation, it is seen that the geometric characteristics are of five types (see ''5-TYPES'' column at right):

1. FORM tolerance—A form tolerance states how far an actual surface or feature is permitted to vary from the desired form implied by the drawing.

2. ORIENTATION tolerance—An orientation tolerance states how far an actual surface or feature is permitted to vary relative to a datum or datums.

3. PROFILE tolerance—A profile tolerance states how far an actual surface or feature is permitted to vary from the desired form on the drawing and/or vary relative to a datum or datums.

4. RUNOUT tolerance—A runout tolerance states how far an actual surface or feature is permitted to vary from the desired form implied by the drawing during full (360°) rotation of the part on a datum axis.

5. LOCATION tolerance—A location tolerance states how far an actual size feature is permitted to vary from the perfect location implied by the drawing as related to a datum, or datums, or other features.

KINDS OF FEATURES TO WHICH A GEOMETRIC CHARACTERISTIC IS APPLICABLE

The geometric characteristics are also divisible into three ''kinds'' of features to which a particular characteristic is applicable. (See ''KIND OF FEATURE'' column at right.)

1. *INDIVIDUAL* feature—A single surface, element, or size feature which relates to a perfect geometric counterpart of itself as the desired form; no datum is proper nor used. (characteristics □, —, ○, ◻).

2. *RELATED* feature—A single surface or element feature which relates to a datum, or datums, in form and orientation. (characteristics ⊥, ∠, //).
    —A size feature (e.g., hole, slot, pin, shaft) which relates to a datum, or datums, in form, attitude (orientation), runout, or location: (characteristics ⊥, ∠, //, ↗, ↗↗, ⊕, ◎).

3. *INDIVIDUAL* or *RELATED* feature—A single surface or element feature whose perfect geometric profile is described which may, or may not, relate to a datum, or datums. (characteristics ⌒, ⌒).

OTHER SYMBOLS AND TERMS

For review, other symbols and characteristics are shown at right. The Projected Tolerance Zone, Least Material Condition, and Datum Target symbols are explained in later text.

GEOMETRIC CHARACTERISTICS, SYMBOLS, AND TERMS

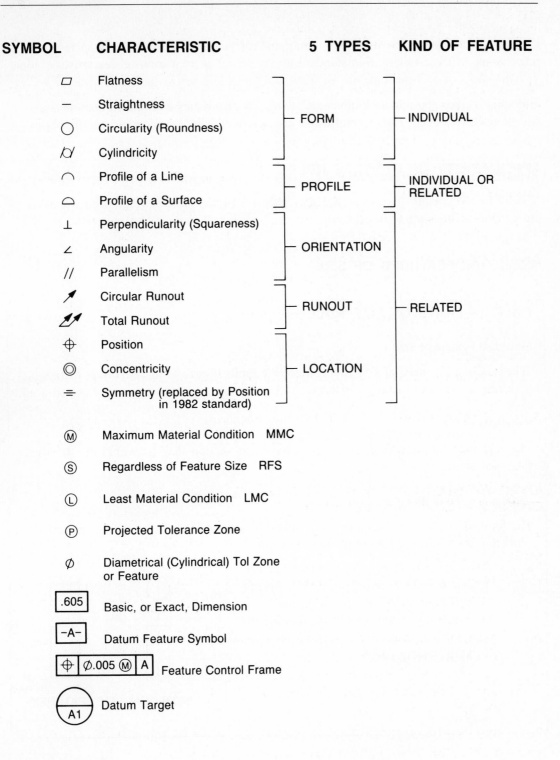

SYMBOL	CHARACTERISTIC	5 TYPES	KIND OF FEATURE
▱	Flatness	FORM	INDIVIDUAL
—	Straightness		
○	Circularity (Roundness)		
⌀	Cylindricity		
⌒	Profile of a Line	PROFILE	INDIVIDUAL OR RELATED
⌓	Profile of a Surface		
⊥	Perpendicularity (Squareness)	ORIENTATION	RELATED
∠	Angularity		
//	Parallelism		
↗	Circular Runout	RUNOUT	
↗↗	Total Runout		
⊕	Position	LOCATION	
◎	Concentricity		
⩮	Symmetry (replaced by Position in 1982 standard)		

Ⓜ Maximum Material Condition   MMC

Ⓢ Regardless of Feature Size   RFS

Ⓛ Least Material Condition   LMC

Ⓟ Projected Tolerance Zone

∅ Diametrical (Cylindrical) Tol Zone or Feature

.605 Basic, or Exact, Dimension

–A– Datum Feature Symbol

⊕ | ∅.005 Ⓜ | A   Feature Control Frame

A1 Datum Target

21

GENERAL RULES

Like any discipline, geometric dimensioning and tolerancing is based on certain fundamental rules. Some of these follow from standard interpretation of the various characteristics, some govern specification, and some are General Rules applying across the entire system.

The various rules appropriate to given geometric characteristics and related nomenclature will be discussed later. The General Rules are described below and on succeeding pages.

APPLICABILITY OF GENERAL RULES

ANSI Y14.5M contains five General Rules. ANSI Y14.5M must be referenced whenever these General Rules are to be applied.

INDIVIDUAL FEATURES OF SIZE

RULE 1—LIMITS OF SIZE RULE

Individual Feature of Size

Where only a tolerance of size is specified, the limits of size of an individual feature prescribe the extent to which variations in its geometric form as well as size are allowed.

Variations of Size

The actual size of an individual feature at any cross-section shall be within the specified tolerance of size.

Variations of Form (Envelope Principle)

The form of an individual feature is controlled by its limits of size to the extent prescribed in the following paragraph and illustration:

a. The surface, or surfaces, of a feature shall not extend beyond a boundary (envelope) of perfect form at MMC. This boundary is the true geometric form represented by the drawing. No variation is permitted if the feature is produced at its MMC limit of size.

INDIVIDUAL SIZE FEATURES

(continued)

(Rule 1 continued)

EXTERNAL FEATURE   **INTERNAL FEATURE**

$\emptyset$.874 $\begin{array}{c}+.000\\-.004\end{array}$ $\emptyset$.875 $\begin{array}{c}+.005\\-.000\end{array}$

$\emptyset$.874 (MMC)

$\emptyset$.880

$\emptyset$.875

$\emptyset$.874 — BOUNDARY OF PERFECT FORM AT MMC

$\emptyset$.870

$\emptyset$.875 (MMC)

$\emptyset$.870

$\emptyset$.874 — BOUNDARY OF PERFECT FORM AT MMC

$\emptyset$.880

$\emptyset$.875

LMC SIZE

b. Where the actual size of a feature has departed from MMC toward LMC, a variation in form is allowed equal to the amount of such departure.

c. There is no requirement for a boundary of perfect form at LMC. Thus, a feature produced at its LMC limit of size is permitted to vary from true form to the maximum variation allowed by the boundary of perfect form at MMC.

WHEN PERFECT FORM AT MMC (RULE 1) DOES *NOT* APPLY:

The control of geometric form prescribed by limits of size does not apply to the following:

a.   Stock such as bars, sheets, tubing, structural shapes, and other items produced to established industry or government standards that prescribe limits for straightness, flatness, and other geometric characteristics. Unless geometric tolerances are specified on the drawing of a part made from these items, standards for the items govern the surfaces that remain in the "as furnished" condition on the finished part.

b.   Parts subject to free state variation in the unrestrained condition.

PERFECT FORM AT MMC *NOT* REQUIRED (RULE 1 REMOVED)

Where it is desirable to permit a surface, or surfaces, of a feature to exceed the boundary of perfect form at MMC, a note such as PERFECT FORM AT MMC NOT REQUIRED is specified exempting the pertinent size dimension from the provision of RULE 1.

RELATIONSHIP BETWEEN INDIVIDUAL FEATURES

The limits of size *do not* control the runout, orientation, or location relationship *between* individual features.

Features shown perpendicular, coaxial, or otherwise geometrically related to each other must be controlled to avoid incomplete drawing requirements. These controls for location, orientation, etc. may be specified by the use of appropriate geometric tolerances.

If it is necessary to establish a boundary of perfect form at MMC to control the relationship between features, the following methods are used:

a.   Specify a zero tolerance of orientation at MMC, including a datum reference (at MMC if applicable), to control perpendicularity, parallelism, or angularity of the feature.

b.   Specify a zero positional tolerance at MMC, including datum references (at MMC if applicable), to control locational, coaxial, or symmetrical features.

c.   Relate the dimensions to a datum reference frame.

d.   Indicate this control for the features involved by a note such as PERFECT ORIENTATION (OR COAXIALITY OR SYMMETRY) AT MMC REQUIRED FOR RELATED FEATURES.

APPLICABILITY OF RFS, MMC, AND LMC

Applicability of RFS, MMC, or LMC is *limited to features subject to variations in size.* They may be datum features or other features whose axes or center planes are controlled by geometric tolerances. In such cases the following practices apply:

┌─RULE 2—POSITION TOLERANCE RULE─────────────

For a tolerance of position, Ⓜ, Ⓢ, or Ⓛ must be specified on the drawing with respect to the individual tolerance, datum reference, or both, as applicable.

| ⊕ | ⌀.005Ⓜ | A | BⓂ |

| ⊕ | ⌀.005Ⓢ | A | BⓈ |

| ⊕ | ⌀.005Ⓛ | A | BⓈ |

┌─FORMER ALTERNATE PRACTICE RULE 2*──────────

Alternate Practice (MMC implied on ⊕ unless specified RFS).
For a tolerance of POSITION (formerly called true position), Ⓜ applies with respect to an individual tolerance, datum reference(s), or both, where no condition is specified. Ⓢ must be specified where it is required.

(IMPLIES Ⓜ ON SIZE FEATURES UNLESS Ⓢ SPECIFIED)

| ⊕ | ⌀.005 | A | B | | ⊕ | ⌀.005Ⓢ | AⓈ | BⓈ |

---

*Formerly recommended practice under USASI Y14.5–1966 and optionally permitted under ANSI Y14.5–1973; retained for explanation purposes on older drawings. This practice *not* permitted under ANSI Y14.5M—1982 and is contrary also to international practice.

┌─RULE 3—OTHER THAN POSITION TOLERANCE RULE ──

For *all* applicable geometric tolerances, other than position tolerance, RFS applies with respect to the individual tolerance, datum reference, or both, where no modifying symbol is specified. Ⓜ must be specified on the drawing where it is required.

─── ⊥ ∠ ∥   (Ⓜ MAY BE SPECIFIED ON SIZE FEATURES AS NECESSARY)

NOTE   The below characteristics and controls are *always* applicable at RFS and due to the nature of the control cannot be applied at MMC.

┌─PITCH DIAMETER RULE◆

Each tolerance of orientation or position and datum reference specified for a screw thread applies to the axis of the thread derived from the pitch cylinder. Where an exception to this practice is necessary, the specific feature of the screw thread (such as MAJOR ⌀ or MINOR ⌀) shall be stated beneath the feature control frame or beneath the datum feature symbol, as applicable.

Each tolerance of orientation or location and datum reference specified for gears, splines, etc. must designate the specific feature of the gear, spline, etc. to which it applies (such as PITCH ⌀, PD, MAJOR ⌀, or MINOR ⌀). This information is stated beneath the feature control frame or beneath the datum feature symbol.

◆Formerly called Rule #4 of ANSI Y14.5–1973. Reference Paragraphs 2.9 and 2.10 of ANSI Y14.5M–1982.

┌─DATUM/VIRTUAL CONDITION RULE■

Depending on whether it is used as a primary, secondary, or tertiary datum, a virtual condition exists for a datum feature of size where its axis or center plane is controlled by a geometric tolerance. In such a case, the datum feature applies at its virtual condition even though it is referenced in a feature control frame at MMC.

■Formerly called Rule #5 of ANSI Y14.5–1973. Reference Paragraph 2.11.1 of ANSI Y14.5M–1982. See DATUM section beginning page 260 for more detail.

Where it is *not* intended for the virtual condition to apply, an appropriate tolerance should be specified to control the form or orientation of the datum feature at its MMC. For additional information see DATUM section and pages 266–271 for examples of application and explanation.

APPLICABILITY OF MMC, RFS, OR LMC

Characteristic	Applicability to feature	Applicability to datum reference
▱ Flatness	Not applicable	No datum reference
— Straightness	MMC Ⓜ or RFS Ⓢ applicable if tolerance applies to axis or center plane of a feature with size; not applicable if considered feature is a single plane surface	
◯ Circularity	Not applicable	
⌀ Cylindricity		
⌒ Profile of a surface ⌒ Profile of a line	Not applicable	RFS on datum reference applicable if datum feature has size and has an axis or center plane; not applicable if datum feature is a single plane surface. MMC Ⓜ not applicable†
⊥ Perpendicularity ∥ Parallelism ∠ Angularity	MMC Ⓜ, RFS Ⓢ, or LMC Ⓛ applicable if tolerance applies to axis or center plane of a feature with size; not applicable if considered feature is a single plane surface	MMC Ⓜ, RFS Ⓢ, or LMC Ⓛ on datum reference applicable if datum feature has size and has an axis or center plane; not applicable if datum feature is a single plane surface
⌖ Position (Symmetry)	MMC Ⓜ, RFS Ⓢ or LMC Ⓛ applicable if tolerance applies to axis or center plane of a feature with size; not applicable if considered feature is a single plane surface	MMC Ⓜ, RFS Ⓢ or LMC Ⓛ on datum reference applicable if datum feature has size and has an axis or center plane; not applicable if datum feature is a single plane surface
↗ Circular Runout ↗↗ Total Runout ◎ Concentricity	Only RFS applicable	Only RFS applicable

†May have exceptions under special conditions. (See PROFILE section)

SHAPE OF TOLERANCE ZONE FOR FORM, ORIENTATION, PROFILE, RUNOUT, AND LOCATIONAL TOLERANCES

Where the specified tolerance value represents the diameter of a cylindrical zone, the diameter symbol ⌀ shall be included in the feature control frame.

Where the tolerance zone is other than a diameter, the tolerance value represents the distance between two parallel straight lines or planes or the distance between two uniform boundaries.

EXAMPLES

TOLERANCE ZONE SHAPE IS: Where the diameter (cylindrical) symbol ⌀ is specified, the tolerance is a diameter (or cylindrical) shape. Where no ⌀ is specified, the tolerance zone is between two parallel lines or planes in the direction of the dimension arrows. The tolerance indicated is the TOTAL tolerance permitted.

POSITION	CONCENTRICITY	SYMMETRY	FLATNESS
⊕ |⌀.XXX Ⓜ| A | —DIAMETER (CYL) TOL ZONE ⊕ |.XXX Ⓜ| A| —TOTAL WIDTH TOL ZONE	◎ |⌀.XXX| A| —DIAMETER (CYL) TOL ZONE	⊕ |.XXX| A| —TOTAL WIDTH TOL ZONE	▱ |.XXX| —TOTAL WIDTH TOL ZONE
STRAIGHTNESS	CIRCULARITY	CYLINDRICITY	PERPENDICULARITY
— |.XXX| —TOTAL WIDTH TOL ZONE — |⌀.XXX Ⓜ| —DIAMETER (CYL) TOL ZONE	○ |.XXX| └TOTAL WIDTH TOL ZONE BETWEEN TWO CONCENTRIC CIRCLES	⌀̸ |.XXX| └TOTAL WIDTH OR THICKNESS TOL ZONE BETWEEN TWO CONCENTRIC CYLINDERS	⊥ |.XXX| A| —TOTAL WIDTH TOL ZONE ⊥ |⌀.XXX| A| —DIAMETER TOL ZONE
PARALLELISM	ANGULARITY	RUNOUT	PROFILE OF SURFACE PROFILE OF LINE
// |.XXX| A| —TOTAL WIDTH TOL ZONE // |⌀.XXX| A| —DIAMETER (CYL) TOL ZONE	∠ |.XXX| A| —TOTAL WIDTH TOL ZONE	↗ |.XXX| A–B| —CIRCULAR RUNOUT (FIM) TOL ZONE ↗↗ |.XXX| A–B| —TOTAL RUNOUT (FIM) TOL ZONE	⌓ |.XXX| A| ⌒ |.XXX| —TOTAL WIDTH OR THICKNESS TOL ZONE AS APPLICABLE

VIRTUAL CONDITION

It is necessary to understand Virtual Condition as it applies to features. The definition below and the examples throughout the text will clarify its meaning.

Definition. The virtual condition of a feature is a derived size generated from the collective effect of all profile variations permitted by the specified tolerances. It represents the most extreme condition of assembly at MMC.

The virtual condition of a feature is the effective size of the profile that must be considered in determining the clearance between mating parts or features.

Size + orientation or position error = shaft virtual condition
Size − orientation or position error = hole virtual condition

EXAMPLE

MEANING

VIRTUAL SIZE, a term also used based on earlier U.S. standard practice, has the same meaning as VIRTUAL CONDITION.

NOTE   See other sections of text and samples for further explanation as applied to controlled features and datum features. See DATUM section for Datum Virtual Condition Rule explanation.

TOLERANCES OF FORM, ORIENTATION, PROFILE, AND RUNOUT

— ▱ ◯ ⌀ = **FORM**

⊥ ∠ ∥ = **ORIENTATION**

⌒ ⌒ = **PROFILE**

↗ ↗↗ = **RUNOUT**

Tolerances of form, orientation, profile, and runout state how far actual surfaces or features are permitted to vary from those implied by the drawing. Expressions of form, orientation, and profile tolerances refer to flatness, straightness, circularity, cylindricity, parallelism, perpendicularity, angularity, profile of a surface, and profile of a line. Runout tolerance is a unique variation and a combination of form and orientation tolerance, and is considered as a separate type of characteristic.

Form, orientation, profile, or runout tolerances should be specified for all features critical to the design requirements:

a.   where established workshop practices cannot be relied upon to provide the required accuracy;

b.   where documents establishing suitable standards of workmanship cannot be prescribed;

c.   where tolerances of size and location do not provide the necessary control.

The various tolerances of form, orientation, profile and runout often have an effect on one another; that is, parallelism could include flatness or straightness, and runout could include circularity, straightness, or cylindricity.

The following series of form, orientation, profile, and runout tolerance examples address each of these characteristics and their individual purpose and meaning, in order to explain the basic principles.

TOLERANCES OF FORM, ORIENTATION, AND PROFILE

Tolerances of form, orientation, and profile provide methods by which to control part geometry where size or location does not adequately do so. Such tolerances state how far an actual surface or feature is permitted to vary from the desired geometric shape implied by the drawing.

KINDS OF FEATURES TO WHICH A FORM, ORIENTATION, OR PROFILE TOLERANCE IS APPLICABLE

To correctly apply form, orientation, or profile tolerances, an understanding of the kind of features, INDIVIDUAL OR RELATED, upon which each characteristic can be used is required. Such tolerances can be applied as follows:

INDIVIDUAL feature—A *single surface, element,* or *size* feature which relates to a perfect geometric counterpart of itself as the desired form; no datum is proper or used.

Form Characteristics which can be applied:

INDIVIDUAL FEATURES

RELATED surface feature—A *single surface or element* feature which relates to a datum, or datums, in orientation.*

Orientation Characteristics which can be applied:

RELATED SURFACE FEATURES

---

*Those tolerances involving related features and datums are sometimes also referred to as attitude tolerances.

RELATED size feature—A single *size* feature which relates to a datum, or datums, in form and orientation.*

Orientation Characteristics which can be applied:

RELATED SIZE FEATURES

INDIVIDUAL or *RELATED* surface or size feature— A *single surface, element* or *size* feature whose perfect geometric profile is described, which may or may not relate to a datum, or datums, in form, orientation, and profile.*

Profile Characteristics which can be applied:

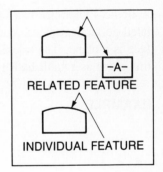

RELATED FEATURE

INDIVIDUAL FEATURE

TOLERANCES OF FORM—INDIVIDUAL FEATURES—NO DATUM

Tolerances of form used on individual features where no datum is proper nor used involve the characteristics below:

(FLATNESS)   (STRAIGHTNESS) (CIRCULARITY) (CYLINDRICITY)

These characteristics are used to describe form tolerances of *single surface, element,* or *size* features and relate to a perfect geometric counterpart of itself.

See following pages for details of application.

---

*Those tolerances involving related features and datums are sometimes also referred to as attitude tolerances.

FLATNESS

Definition. Flatness is the condition of a surface having all elements in one plane.

FLATNESS TOLERANCE

Flatness tolerance specifies a tolerance zone confined by two parallel planes within which the entire surface must lie.

FLATNESS TOLERANCE APPLICATION

The example below shows how a flatness symbol is applied.

The symbol is interpreted to read "This surface shall be flat within .002 total tolerance zone over entire surface." Note that the .002 tolerance zone is *total* variation. To be acceptable, the entire actual surface must fall within the parallel plane extremities of the .002 tolerance zone.

A flatness tolerance is a form control of all elements of a surface as it compares to a simulated perfect geometric counterpart of itself. The perfect geometric counterpart of a flat surface is a plane. The tolerance zone is established as a width or thickness zone relative to this plane as established from the actual part surface.

EXAMPLE

MEANING

WITHIN .002 WIDE TOL ZONE TOTAL OVER ENTIRE SURFACE THIS SURFACE MUST BE FLAT

.002 TOL ZONE

ENTIRE SURFACE MUST LIE BETWEEN THE TWO PARALLEL PLANES

LIMITS OF FORM (PARALLEL)

.002 TOLERANCE ZONE

.002 TOLERANCE ZONE

.655 HIGH SIZE LIMIT

.645 LOW SIZE LIMIT

POSSIBLE VARIATION OF FLATNESS TOL ZONE WITHIN SIZE TOL RANGE

Note that the extremities or high points of the surface determine one limit or plane of the tolerance zone, with the other limit or plane being established .002 (the specified tolerance) parallel to it.

Since flatness tolerancing control is essentially a relationship of a feature to itself, no datum references are required or proper.

Also, note that since flatness is a form tolerance controlling surface elements only, it is not applicable to RFS or MMC considerations.

In the absence of a flatness tolerance specification, the size tolerance and method of manufacture of a part will exercise some control over its flatness. However, when a flatness tolerance is specified, as applicable to a single surface, the flatness tolerance zone must be contained *within* the size tolerance limits.* It cannot be additive to the size tolerance. Therefore, a flatness tolerance should normally be *less than* the part size tolerance.

Where necessary the terms "MUST NOT BE CONCAVE" or "MUST NOT BE CONVEX" may be added beneath the feature control frame.

STRAIGHTNESS —

Definition. Straightness is a condition where an element of a surface or an axis is a straight line.

STRAIGHTNESS TOLERANCE

A straightness tolerance specifies a tolerance zone within which the considered element or axis must lie.

STRAIGHTNESS TOLERANCE APPLICATION

A straightness tolerance is applied in the view where the elements to be controlled are represented by a straight line.

STRAIGHTNESS TOLERANCE—SURFACE ELEMENT CONTROL

Straightness tolerance is typically used as a form control of individual surface elements such as those on cylindrical or conical surfaces. Since surfaces of this kind are made up of an infinite number of longitudinal elements, a straightness requirement applies to the entire surface as controlled in single line elements in the direction specified.

The example below illustrates straightness control of individual longitudinal surface elements on a cylindrical part. Note that the symbol is directed to the feature surface (or extension

---

*Under special circumstances, a note specifically exempting the pertinent size dimension, i.e., PERFECT FORM AT MMC NOT REQUIRED, may be specified.

line) and not to the dimension lines. The straightness tolerance must be less than the size tolerance.

All circular elements of the surface must be within the specified size tolerance and the boundary of perfect form at MMC. Also, each longitudinal element of the surface must lie in a tolerance zone defined by two parallel lines spaced apart by the amount of the prescribed tolerance where the two lines and the nominal axis of the part share a common plane.

NOTE   Since surface element control is specified, the tolerance zone applies uniformly whether the part is of a bowed, waisted, or barreled shape.

EXAMPLE

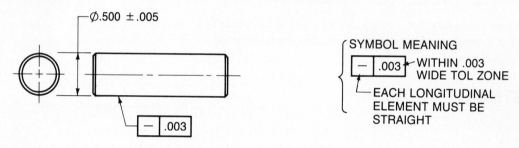

SYMBOL MEANING

| − | .003 | ← WITHIN .003 WIDE TOL ZONE

└ EACH LONGITUDINAL ELEMENT MUST BE STRAIGHT

MEANING

THE FEATURE MUST BE WITHIN THE SPECIFIED TOL OF SIZE AND THE BOUNDARY OF PERFECT FORM AT MMC (.505). EACH LONGITUDINAL ELEMENT OF THE SURFACE MUST LIE BETWEEN TWO PARALLEL LINES (.003 APART) WHERE THE TWO LINES AND THE NOMINAL AXIS OF THE PART SHARE A COMMON PLANE.

*BOUNDARY OF PERFECT FORM AT MMC

Note the absence of any datum reference when straightness tolerancing is used. Straightness is a form control of a single element, or an axis, as it relates to a perfect geometric counterpart of itself. Therefore, datum references are neither required nor proper.

STRAIGHTNESS TOLERANCE APPLIED TO FLAT SURFACE

Straightness tolerancing may be applied to flat surfaces to provide surface element control in a specific direction as a refinement of size tolerance or other form tolerance such as flatness. If so used, the straightness tolerance must be less than the refined tolerance.

The example below illustrates straightness tolerance of surface elements as a refinement of size tolerance. Note that the tolerance is applied in the view in which the elements to be controlled appear as a straight line.

The individual straightness elements must be within both the size tolerance and the straightness tolerance zone of .003, whereas element to element in the other view, variation within the size tolerance may occur.

EXAMPLE

SYMBOL MEANING

| — | .003 | ← WITHIN .003 WIDE TOL ZONE

EACH ELEMENT (IN DIRECTION SHOWN) MUST BE STRAIGHT

| — | .003 |

1.000 ± .005

MEANING

.003 TOL ZONE

SIZE TOL

.003

.003

STRAIGHTNESS TOLERANCE RFS AND MMC

Where function of a size feature permits a collective result of *size and form* variation known as the virtual condition,* the RFS or MMC principles may be used.

In this instance, where the appropriate symbology and specifications are used, the part is not confined to the perfect form at MMC boundary. All sectional elements of the surface are to be within the specified size tolerance, but the total part surface may exceed the perfect form at MMC boundary to the extent of the straightness tolerance.

This principle may be applied to individual size features such as pins, shafts, bars, etc., where the longitudinal elements are to be specified with a straightness tolerance independent of, or in addition to, the size tolerance.

STRAIGHTNESS—RFS—VIRTUAL CONDITION

Where a cylindrical feature is to be controlled on an RFS basis, as below, the feature control symbol must be located with the size dimension or attached to the dimension line, and the diameter symbol must precede the straightness tolerance.

EXAMPLE

SYMBOL MEANING

—OF .015, RFS

WITHIN ⌀ (CYL) TOL ZONE

THE AXIS MUST BE STRAIGHT

MEANING

EACH CIRCULAR ELEMENT OF THE FEATURE MUST BE WITHIN THE SPECIFIED TOLERANCE OF SIZE. THE DERIVED AXIS OF THE ACTUAL FEATURE MUST LIE WITHIN A DIA OR CYLINDRICAL TOLERANCE ZONE .015 REGARDLESS OF FEATURE SIZE.

---

*VIRTUAL CONDITION—Virtual condition of a feature is the collective effect of size and form error that must be considered in determining the fit or clearance between mating parts or features. See above example.

STRAIGHTNESS—MMC—VIRTUAL CONDITION

Where a cylindrical feature has a functional relationship with another feature, such as a pin or shaft and a hole, the control of straightness on an MMC basis may be desirable. If the pin or shaft, for example, is to fit into a hole of a given diameter, the collective effect of the pin size and its straightness error must be considered in relationship to the hole size minimum, i.e., their virtual conditions must be considered relative to one another.

On the part below, the size must be maintained at all sectional elements within stated limits. Likewise its straightness, using the axis as the criteria, must be within tolerance, but only when the part is at MMC. Therefore, the part develops (or is based upon) a virtual condition of .517, which represents the extreme condition the part can have and yet perform its function, or fit the mating part.

By stating the requirements on an MMC basis, the allowable straightness tolerance may increase an amount equal to the size departure from MMC. The feature control symbol must be located with the size dimension, or attached to the dimension line; the diameter symbol must precede the straightness tolerance; and the MMC symbol must be inserted following the tolerance. In this manner maximum tolerance is achieved, part fit is guaranteed, and functional gaging techniques may be used.

EXAMPLE

MEANING

IF FEATURE SIZE ∅	∅ TOL ZONE IS
.502 MMC	.015
.501	.016
.500	.017
.499	.018
.498 LMC	.019

EACH CIRCULAR ELEMENT OF THE FEATURE MUST BE WITHIN THE SPECIFIED TOLERANCE OF SIZE. THE DERIVED AXIS OF THE ACTUAL FEATURE MUST LIE WITHIN A DIA OR CYLINDRICAL TOLERANCE ZONE OF .015 AT MMC. AS THE FEATURE DEPARTS FROM MMC, AN INCREASE IN THE STRAIGHTNESS TOLERANCE IS ALLOWED WHICH IS EQUAL TO THE AMOUNT OF SUCH DEPARTURE.

Where straightness tolerance on an MMC basis is specified, functional gaging techniques may be used. The below gage and conditions demonstrate how these principles can be applied to the preceding part.

GAGE

CONDITIONS

∅.517 VIRTUAL CONDITION POSSIBLE

$$\begin{pmatrix} \text{MMC SIZE} & .502 \\ + \text{STR TOL} & .015 \\ \hline & .517 \end{pmatrix}$$

∅.502   PART AT HI LIMIT SIZE (MMC) PERFECT STRAIGHTNESS

∅.517 VIRTUAL CONDITION POSSIBLE

∅.015 STR TOL ZONE

∅.502 PART AT MMC SIZE OUT OF STRAIGHT, FULL AMOUNT OF ALLOWABLE TOLERANCE .015

∅.517 VIRTUAL CONDITION POSSIBLE

∅.019 STR TOL ZONE

∅.498 PART AT LMC SIZE PERMITS .019 STRAIGHTNESS:

MMC SIZE	.502	
(−) LMC SIZE	.498	
ADDED TOL	.004 ————	.004
	STR TOL	.015
	TOTAL TOL	.019

From the above it is seen that straightness applied on an MMC basis provides control of mating part conditions to facilitate design and provides maximum production tolerance. Functional gaging techniques are permissible. Note that the above gage simulates both the extreme permissible condition of the part as well as the condition of the mating part.

STRAIGHTNESS ON UNIT LENGTH BASIS

Where required, straightness may be applied on a unit length basis. This method is occasionally used to facilitate special design requirements and to prevent abrupt surface variations within a relatively short length of the feature. To prevent extreme variations of bow over the total length of the part, the amount of permissible straightness variation allowable on the total length of the part should be specified. If the unit variations are permitted to continue along the length of the part with no maximum tolerance indicated, an unsatisfactory part could result.

The example below illustrates a part with unit straightness specified on an RFS basis (MMC principles could also be used if desired.)

EXAMPLE

SYMBOL MEANING

THE AXIS MUST BE STRAIGHT

WITHIN ∅ (CYL) TOL ZONE

OF .020, RFS TOTAL LENGTH

AND WITHIN ∅ (CYL) TOL ZONES OF .005, .020 RFS

FOR EACH 1.000 IN. UNIT OF LENGTH

MEANING

EACH CIRCULAR ELEMENT OF THE FEATURE MUST BE WITHIN THE SPECIFIED TOLERANCE OF SIZE. THE DERIVED AXIS OF THE ACTUAL FEATURE MUST LIE WITHIN A CYLINDRICAL TOLERANCE ZONE OF .020 FOR THE TOTAL 4.00 LENGTH, AND WITHIN A .005 CYLINDRICAL TOLERANCE ZONE FOR EACH 1.000 LENGTH, RFS.

CIRCULARITY ◯

Definition. Circularity is the condition on a surface of revolution where:

1.   in the case of a cylinder or cone, all points of the surface intersected by any plane perpendicular to a common axis are equidistant from their axis;

2.   in the case of a sphere, all points of the surface intersected by any plane passing through a common center are equidistant from that center.

CIRCULARITY TOLERANCE

A circularity tolerance specifies a tolerance zone bounded by two concentric circles within which each circular element of the surface must lie and applies independently at any plane as described above.

CIRCULARITY TOLERANCE APPLICATION

Limits of size exercise control of circularity within the size tolerance. Often this provides adequate control. However, where necessary to further refine form control, circularity tolerancing can be used on any figure of revolution or circular cross section.

The example illustrates a part with a circularity tolerance of .002 specified on a cylindrical part.

The interpretation shows how one establishes the .002 tolerance zone. Note that the tolerance zone is the width of the annular zone between the two concentric circles.

A circularity tolerance zone is established relative to the actual size of the part when measured at the surface periphery at any cross section perpendicular to the part axis. It should be noted that the circularity tolerance applies only at the cross-sectional point of measurement, and is relative to the *size* at that point. Therefore, a cylindrical part with circularity tolerance control could taper or otherwise vary in its surface contour within its size tolerance range, yet still meet circularity requirements if it is within the circularity tolerance at that point.

The part size in this example has been assumed to measure .503 at its largest point at the cross section selected for measurement. The .002 circularity tolerance zone is then established by two theoretically perfect concentric circles, one at the .503 diameter and the other .004 *smaller* at the .499 diameter. This establishes the tolerance zone of .002 *width* between the concentric circles. To be acceptable, the part surface at that cross section must fall within the .002 wide tolerance zone.

As is seen, the tolerance zone is established relative to the part size wherever it may fall in its size tolerance range. That is, the part *size* is first determined and its circularity is then defined as a refinement of the part *form* relative to that *size*. Unless otherwise specified, any established size at any point along the surface can be used to determine the circularity tolerance zone. It is therefore seen that the circularity tolerance may be based on *different* sizes on the same part. The circularity tolerance zone, however, remains constant.

Note again that the circularity tolerance must always be contained *within* the part *size* tolerance range. The circularity tolerance cannot exceed the size tolerance limits. Furthermore, a circularity tolerance cannot be modified to an MMC application since it controls surface elements only.

A circularity tolerance is a form control of a single part element as it compares to a perfect counterpart of itself. Therefore, no datum references are required nor proper. A circularity tolerance does have a reference center, but this is considered only as a part of the perfect frame of reference (concentric circles and their common center) for measurement, just as a straightness tolerance refers to a perfect line of reference for its measurement. You may compare a circularity tolerance to a straightness tolerance zone curled around a circle.

CIRCULARITY OF A CYLINDER

EXAMPLE   **MEANING**

SECT A-A

SURFACE PERIPHERY AT ANY CROSS SECTION PERPENDICULAR TO THE AXIS
MUST BE WITHIN THE SPECIFIED TOLERANCE OF SIZE AND MUST LIE
BETWEEN TWO CONCENTRIC CIRCLES, ONE HAVING A RADIUS .002 LARGER
THAN THE OTHER.

(ABOVE SIZES ARBITRARILY SELECTED FOR ILLUSTRATION)

CIRCULARITY OF A CONE

Example 1 below illustrates a cone-shaped part for which a circularity tolerance of .003 is specified. As previously discussed, the periphery at any cross section perpendicular to the axis must be within the specified tolerance of size and must lie between the two concentric circles (one having a radius .003 larger than the other).

EXAMPLE 1

Ø.850 ±.005
Ø.405 ±.005

⊙ .003

MEANING

90°

A

A

.003 WIDE TOLERANCE ZONE

SECTION A-A

SYMBOL MEANING

⊙ .003

WITHIN .003 WIDE TOL ZONE

THIS FEATURE MUST BE CIRCULAR

THE PERIPHERY AT ANY CROSS SECTION PERPENDICULAR TO THE AXIS MUST BE WITHIN THE SPECIFIED TOLERANCE OF SIZE AND MUST LIE BETWEEN TWO CONCENTRIC CIRCLES (ONE HAVING A RADIUS .003 LARGER THAN THE OTHER).

CIRCULARITY OF A SPHERE

Circularity of a spherical part is given the same basic interpretation (see Example 2 below) except that the tolerance control reference is to *any cross section passing through a common center* rather than to *any cross section perpendicular to the axis*, as in the conventional application of circularity tolerancing.

EXAMPLE 2

S⌀.500 ±.005
| ◯ | .003 |

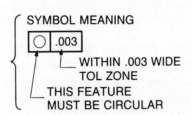

SYMBOL MEANING

| ◯ | .003 |

WITHIN .003 WIDE
TOL ZONE
THIS FEATURE
MUST BE CIRCULAR

THE PERIPHERY AT ANY CROSS SECTION PASSING THROUGH A COMMON CENTER MUST BE WITHIN THE SPECIFIED TOLERANCE OF SIZE AND MUST BE BETWEEN TWO CONCENTRIC CIRCLES (ONE HAVING A RADIUS .003 LARGER THAN THE OTHER). HENCE, THE SURFACE MUST LIE BETWEEN TWO CONCENTRIC SPHERES SEPARATED .003 APART.

MEANING

SECTION A- A

.003 WIDE TOLERANCE
ZONE

CYLINDRICITY ⌭

Definition. Cylindricity is the condition of a surface of revolution in which all points (elements) of the surface are equidistant from a common axis.

CYLINDRICITY TOLERANCE

A cylindricity tolerance specifies a tolerance zone bounded by two concentric cylinders within which the surface must lie.

CYLINDRICITY TOLERANCE APPLICATION

Limits of size exercise control of cylindricity within the size tolerance. This control is often adequate. However, where more refined form control is required, cylindricity tolerancing can be used. Note that in cylindricity, unlike circularity, the tolerance applies simultaneously to both circular and longitudinal elements of the entire surface.

The example illustrates a part with a cylindricity tolerance of .003. A cylindricity tolerance is interpreted as "on the radius" or, as in this case, within the .003 wide tolerance zone defined by two concentric cylinders .003 apart. A cylindricity tolerance can be considered circularity tolerancing extended to control the *entire* surface of a cylinder.

EXAMPLE

Ø.500 ±.005

⌭ .003

SYMBOL MEANING

⌭ .003

— WITHIN .003 WIDE TOL ZONE
— THIS FEATURE MUST BE CYLINDRICAL

THE FEATURE MUST BE WITHIN THE SPECIFIED TOLERANCE OF SIZE AND MUST LIE BETWEEN TWO CONCENTRIC CYLINDERS (ONE HAVING A RADIUS .003 LARGER THAN THE OTHER).

MEANING

.003   TOLERANCE ZONE

.497   .503

46

Cylindricity tolerancing can be applied *only* to cylindrical forms. The leader from the feature control symbol may be directed to either view.

It should be noted that a cylindricity tolerance simultaneously controls circularity, straightness, and parallelism of the elements of the cylindrical surface.

A cylindricity tolerance applies to the *entire* cylindrical surface as opposed to the cross-sectional or diametral measurement considered in circularity. Also, in measuring cylindricity, the concentric cylinders defining the tolerance zone are always based on the size of the part produced. The part size in the example has been assumed to measure .503 at its largest diameter. This .503 is the diameter of the largest of the concentric cylinders defining the tolerance zone. The smaller of the concentric cylinders is .503 minus the amount of the cylindricity tolerance (.003 on R = .006 on dia), i.e., .497. The cylindricity tolerance zone is therefore .003 *wide* between the concentric cylinders. The entire part surface must then fall within this tolerance zone to be acceptable.

It should be noted that the cylindricity tolerance must always be contained *within* the part *size* tolerance range. The cylindricity tolerance cannot exceed the size tolerance limits.

As is seen, the cylindricity tolerance zone is established relative to the part size wherever it may fall in its *size* tolerance range. That is, the part size is first determined and its *cylindricity* is then defined as a refinement of the part *form* relative to that *size*.

If the largest measurement of the produced part had been near the *low* limit, for example at .497, the cylindricity tolerance could not have been more than .001 on R (= .002 on dia). The cylindricity tolerance cannot exceed *size* tolerance limits. Therefore, the form (cylindricity) variation could not be less than the low limit size of .495.

A cylindricity tolerance (similar to flatness) is a form control of a surface element as it compares to a perfect counterpart of itself. Therefore, no datum references are required nor proper. One can compare a cylindricity tolerance zone to a flatness tolerance zone by visualizing that flatness zone curled around a cylinder. A cylindricity tolerance does have a reference axis, but this is considered only a part of the perfect frame of reference (concentric cylinders and their common axis) for measurement, just as a flatness tolerance refers to a perfect plane of reference for its measurement.

Since cylindricity is a form tolerance controlling surface elements only, it cannot be modified to an MMC application.

EVALUATION OF CIRCULARITY AND CYLINDRICITY

Cylindricity is checked or measured by the same basic techniques that are used to check circularity except that cylindricity involves a cylindrical tolerance zone of uniform thickness over the entire surface and is based on a single size reference. This must be considered in any measuring procedure concerned with cylindricity. For example, in the following discussions of vee block, between centers, and polar graph methods, the measurements are made at cross sections only; whereas in cylindricity, one must consider the entire surface as controlled by the one tolerance zone.

Vee block or between centers methods are often used in open set-up inspection of circularity and cylindricity.

The vee block method provides an approximation or a rough check which may be adequate for some applications. Variables, such as the varying number and arrangement of lobes on the part surface and the angle of the vee block, may affect the resultant measurement sufficiently for the method to become rather inaccurate. In addition, the procedure itself contributes to inaccuracies, because one portion of the part surface (in the vee) is used as the basis for checking another portion of the same surface, thus compounding the chances for error. With certain lobe and vee block combinations, a visibly out-of-circularity part can be rotated with no evidence of error on the dial indicator (see next page).

When the vee block method is used, the total error read is roughly the error on the *diameter;* it must be halved, as an approximation, to compare the result with the specified circularity tolerance which is implied as the radial separation or total width between annular circles.

The between centers method more nearly establishes the reference axis for the geometric relationship. Any error in the centers, however, will affect the accuracy of the resultant measurement. Since the part is rotated about its nominal center or axis, the reading will be on the radius and representative of the total width zone between concentric circles. This method, however, is more correctly termed a runout relationship of a surface and centers and is technically not a circularity analysis.

Figures 1 and 2 on the following page demonstrate that certain vee block and lobed-part conditions do not register the correct part circularity or cylindricity error.

Obviously these parts have noticeable form errors. However, when the parts are rotated in the 60° vee block, the true error will not register. In fact, in these hypothetical examples of a five-lobed part and an oval shape, *no* error registers in any position. When the number of lobes of the part is known, certain vee block angles can be used to obtain accurate diametral readings. However, the above indicated variables and the difficulty of accurately predicting conditions on any given part make this method impractical.

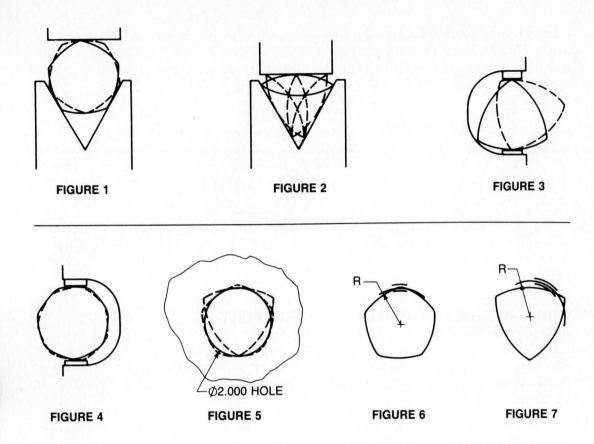

FIGURE 1   FIGURE 2   FIGURE 3

FIGURE 4   FIGURE 5   FIGURE 6   FIGURE 7

Figures 3 and 4 illustrate how "miking" a lobed part (particularly a part with an uneven number of lobes) will not pick up error. The parts shown, although obviously not true in form, register the same measurements at any diametrical location. Suppose the parts in Fig. 1 (five lobes) and Fig. 3 (three lobes) were "miked" at 1.999 since they are intended to go through the 2.000 ϕ hole shown in Fig. 5. The parts will not pass through the hole although they were "miked" at the lower size measurement of 1.999.

Figures 6 and 7 illustrate the parts shown in Figs. 1 and 3 evaluated on the basis of radial values. Using appropriate measuring techniques, the error can be determined directly.

PRECISION METHOD OF EVALUATING CIRCULARITY AND CYLINDRICITY

To evaluate circularity and cylindricity precisely, one must relate the part surface periphery to the geometry of a perfectly round or cylindrical form as constructed from a reference axis. Several kinds of special gaging equipment utilizing optical, mechanical, electronic, and pneumatic principles are available. One method uses an electronic probe which travels around the periphery of the part while the part is chuck-mounted on an extremely accurate spindle and transcribes an enlarged profile of the part periphery on a polar graph. This profile is then compared with a transparent overlay gage which contains circles at various increments. Note that the final basis for comparison is the part profile only and the reference axis is merely a means of constructing the geometry for measurement. Although more costly and time consuming than other methods, this method utilizes geometric relationships which more directly evaluate circularity and cylindricity where necessary.

OTHER METHODS OF EVALUATING CIRCULARITY AND CYLINDRICITY

Due to the increasing sophistication of design requirements, manufacturing processes, and measuring processes, such control as circularity and cylindricity may be accomplished by yet further methods.

Such methods as "Least Mean Square," which determines the center of a circular form by mathematical formula,

$$X = \frac{2\Sigma X}{N}, \ Y = \frac{2\Sigma Y}{N}, \ R = \frac{\Sigma R}{N},$$

or a computer program based on ordinate and radius measurements may be used. Other methods, such as "minimum circumscribed circle," "maximum inscribed circle," and "minimum radial separation," which utilize precision spindle techniques can be used in appropriate circumstances.

More specific callout of the drawing requirement indicating one of the above methods (or stylus tip radius, cycles per revolution, etc.) may occasionally be necessary.

DATUMS
BASIS FOR RELATING FEATURES TO ONE ANOTHER

MEANING OF DATUM

Flatness, straightness, circularity, cylindricity (and occasionally profile) tolerancing are characteristics or controls which are applied to single features. It might be said that in order to define these controls, the feature is compared to a perfect counterpart of itself with a stated tolerance to indicate the variation permissible from perfect for that feature. It might also be said that the feature serves as its own "datum"; a datum being a feature from which relationships originate. The following definition will further clarify the meaning and intention for a datum.

Definition. A datum is a theoretically exact point, axis, or plane derived from the true geometric counterpart of a specified datum feature. A datum is the origin from which the location or geometric characteristics of features of a part are established.

DATUM VS. DATUM SURFACE AND DATUM FEATURE

A distinction must be made between a "datum" (theoretically exact—as on a drawing), which represents design requirement and function, and the actual "datum surface" or "datum feature" on the produced part.

Definition. A datum feature is an actual feature of a part which is used to establish a datum.

DATUMS REQUIRED TO RELATE FEATURES

When relating form, orientation, profile, runout, or location tolerances to features in relation to one another, a datum (or datums), must be used. Datum precedence also must be considered when necessary. (See below and Datum section.)

Further examples of datum application using form, orientation, profile, runout, and location characteristics are given in the DATUM section and in following examples throughout this text.

ORIENTATION TOLERANCES—RELATED FEATURES USING DATUMS

Orientation tolerances used on related features and which require a datum involve the characteristics below:

(PERPENDICULARITY)   (ANGULARITY)   (PARALLELISM)

These characteristics are used to describe orientation tolerances of *single surface, element,* or *size* features and are always related to a *datum*.

Orientation tolerance when applied to a plane surface must be *less than* the governing *size* tolerance of the features involved. Orientation tolerance when applied to size feature (hole or pin) is a refinement of orientation (perpendicularity, etc.) within the stated locational tolerance; therefore, it must be *less than* the governing locational tolerance of the features involved.

See following pages for details of application.

PERPENDICULARITY ⊥
(SQUARENESS, NORMALITY)

Definition. Perpendicularity is the condition of a surface, median plane, or axis which is at exactly 90° to a datum plane or axis.

PERPENDICULARITY TOLERANCE

A perpendicularity tolerance specifies:

1. a tolerance zone defined by two parallel planes perpendicular to a datum plane within which

   a. the surface of a feature must lie (see Fig. 1);

   b. the median plane of a feature must lie (see Fig. 2);

2. a tolerance zone defined by two parallel planes perpendicular to a datum axis within which the axis of a feature must lie (see Fig. 3);

3. a cylindrical tolerance zone perpendicular to a datum plane within which the axis of a feature must lie (see Fig. 4);

4. a tolerance zone defined by two parallel, straight lines perpendicular to a datum plane or datum axis within which an element of the surface must lie (see Fig. 5—radial perpendicularity).

PERPENDICULARITY APPLICATION

The example below illustrates perpendicularity tolerance as applied to a surface.

EXAMPLE

Note that the perpendicularity tolerance applied to a plane surface controls flatness if a flatness tolerance is not specified (that is, the flatness will be at least as good as the perpendicularity).

The examples below and on the following pages show perpendicularity tolerancing under various conditions. Note also how MMC applications permit greater tolerance.

NONCYLINDRICAL FEATURE AT RFS, DATUM A PLANE

EXAMPLE

NONCYLINDRICAL FEATURE AT RFS, DATUM A PLANE

SYMBOL MEANING

⊥ | .005 | A — TO DATUM PLANE A
— WITHIN .005 WIDE TOL ZONE, RFS
— THIS FEATURE MUST BE PERPENDICULAR

MEANING

POSSIBLE POSITION OF THE FEATURE MEDIAN PLANE

DATUM PLANE

.005 WIDE TOL ZONE

THE FEATURE MEDIAN PLANE MUST BE WITHIN THE SPECIFIED TOLERANCE OF LOCATION. REGARDLESS OF THE ACTUAL SIZE OF THE FEATURE, ITS MEDIAN PLANE MUST LIE BETWEEN TWO PARALLEL PLANES (.005 APART) WHICH ARE PERPENDICULAR TO THE DATUM PLANE.

NONCYLINDRICAL FEATURE AT MMC, DATUM A PLANE

EXAMPLE

NONCYLINDRICAL FEATURE AT MMC, DATUM A PLANE

SYMBOL MEANING

⊥ | .005 Ⓜ | A — TO DATUM PLANE A
— WITHIN .005 WIDE TOL ZONE AT MMC
— THIS FEATURE MUST BE PERPENDICULAR

MEANING

POSSIBLE DIRECTION FOR THE FEATURE MEDIAN PLANE

DATUM PLANE

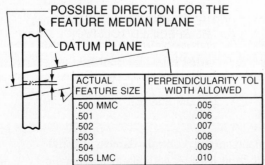

ACTUAL FEATURE SIZE	PERPENDICULARITY TOL WIDTH ALLOWED
.500 MMC	.005
.501	.006
.502	.007
.503	.008
.504	.009
.505 LMC	.010

THE FEATURE MEDIAN PLANE MUST BE WITHIN THE SPECIFIED TOLERANCE OF LOCATION. WHEN THE FEATURE IS AT MAXIMUM MATERIAL CONDITION (.500) THE MAXIMUM PERPENDICULARITY TOLERANCE IS .005 WIDE. WHERE THE FEATURE IS LARGER THAN ITS SPECIFIED MINIMUM SIZE, AN INCREASE IN THE PERPENDICULARITY TOLERANCE IS ALLOWED.

CYLINDRICAL FEATURE AT RFS, DATUM A CYLINDER RFS

EXAMPLE

SYMBOL MEANING

⊥ | .005 | A ← TO DATUM AXIS A, RFS

└ WITHIN .005 WIDE TOL ZONE, RFS

└ THIS FEATURE MUST BE PERPENDICULAR

MEANING

.005 WIDE TOL ZONE

DATUM AXIS

POSSIBLE DIRECTION FOR THE FEATURE AXIS

THE FEATURE AXIS MUST BE WITHIN THE SPECIFIED TOLERANCE OF LOCATION AND MUST LIE BETWEEN TWO PLANES (.005 APART) WHICH ARE PERPENDICULAR TO THE DATUM AXIS.

NOTE: This tolerance applies only to the view on which it is specified.

CYLINDRICAL FEATURE AT MMC, DATUM A PLANE

Note that the ⌀ symbol is required to indicate a diameter (cylindrical) tolerance zone.

EXAMPLE

⌀.250 +.000 −.001

⊥ | ⌀.003 Ⓜ | A

SYMBOL MEANING

⊥ | ⌀.003 Ⓜ | A ← TO DATUM PLANE A

└ WITHIN ⌀.003 TOL ZONE AT MMC

└ THIS FEATURE MUST BE PERPENDICULAR

MEANING

ACTUAL FEATURE SIZE	PERP TOL ⌀ DIA ALLOWED
.250 MMC	.003
.2495	.0035
.249 LMC	.004

.253

FUNCTIONAL GAGE

PART   FLUSH

DATUM PLANE

THE FEATURE MUST BE WITHIN THE SPECIFIED TOLERANCE OF LOCATION. WHEN THE FEATURE IS AT MMC .250 THE PERPENDICULARITY TOLERANCE IS .003 DIAMETER. AS THE FEATURE SIZE DEPARTS FROM MMC (GETS SMALLER) AN INCREASE IN TOLERANCE IS PERMITTED EQUAL TO THE AMOUNT OF THAT DEPARTURE.

CYLINDRICAL FEATURE AT MMC, DATUM A PLANE

Note that the $\emptyset$ symbol is required to indicate a diameter (cylindrical) tolerance zone.

EXAMPLE

SYMBOL MEANING

⊥ | Ø.010 Ⓜ | A ← TO DATUM PLANE A

└ WITHIN Ø.010 TOL ZONE AT MMC

└ THIS FEATURE MUST BE PERPENDICULAR

Ø2.000 $\begin{array}{l}+.005\\-.000\end{array}$

⊥ | Ø.010 Ⓜ | A

MEANING

POSSIBLE DIRECTION FOR THE FEATURE AXIS

ACTUAL FEATURE SIZE	PERPENDICULARITY TOLERANCE DIAMETER Ø ALLOWED
2.000 MMC	.010
2.001	.011
2.002	.012
2.003	.013
2.004	.014
2.005 LMC	.015

DATUM PLANE

WHEN THE FEATURE IS AT MMC (2.000), THE MAXIMUM PERPENDICULARITY TOLERANCE IS .010 DIAMETER. AS THE FEATURE SIZE DEPARTS FROM MMC (GETS LARGER), AN INCREASE IN TOLERANCE IS PERMITTED EQUAL TO THE AMOUNT OF THAT DEPARTURE.

RADIAL PERPENDICULARITY

EXAMPLE

⊥ | .001 | A

EACH RADIAL ELEMENT

-A-

SYMBOL MEANING

⊥ | .001 | A ← TO DATUM AXIS A, RFS

└ WITHIN .001 WIDE TOL ZONE

└ EACH RADIAL ELEMENT MUST BE PERPENDICULAR

MEANING

TOLERANCE ZONE .001

TRAVEL OF THE INDICATOR IS IN A RADIAL DIRECTION WITH PART HELD STATIONARY. EACH RADIAL ELEMENT OF THE SURFACE MUST BE WITHIN THE SPECIFIED TOLERANCE OF SIZE AND MUST LIE BETWEEN TWO PARALLEL LINES (.001 APART) WHICH ARE PERPENDICULAR TO THE AXIS OF DIAMETER A.

Where "element" control is required and the perfect form at MMC requirement of Rule 1 is not applicable, the method below may be used.

EXAMPLE

⚠ PERFECT FORM AT MMC NOT REQUIRED

MEANING

Each perpendicular element of the surface at any location along the entire length of the surface must lie between two parallel lines (.005 apart) which are perpendicular to datum plane A. Each longitudinal element of the surface must lie between two parallel straight lines (.040 apart). The part perfect form of the MMC envelope may be exceeded but the stated tolerances are the maximum form error permissible on the surface.

When perpendicularity tolerancing is critical, it may be necessary to limit the *tolerance* deviation to an amount equal to the feature *size* deviation from MMC. This assumes that the part form must be perfect at MMC size and that the virtual condition (size) can be no greater than that at MMC. The only permissible form tolerance must be acquired from the variation in part size (see Example 1 below) in the increase of the feature hole size.

Example 2 limits the tolerance acquired from the feature size increase to a MAX amount.

ZERO TOLERANCE AT MAXIMUM MATERIAL CONDITION

Note that the ⌀ symbol is required to indicate a diameter (cylindrical) tolerance zone.

EXAMPLE 1

MEANING

ACTUAL FEATURE SIZE	PERPENDICULARITY TOLERANCE DIAMETER ⌀ ALLOWED
2.000 MMC	.000
2.001	.001
2.002	.002
2.003	.003
2.004	.004
2.005 LMC	.005

WHEN THE FEATURE IS AT MMC 2.000 ITS AXIS MUST BE PERPENDICULAR TO THE DATUM PLANE. AS THE FEATURE SIZE DEPARTS FROM MMC (GETS LARGER) AN INCREASE IN TOL IS PERMITTED EQUAL TO THE AMOUNT OF THAT DEPARTURE.

ZERO TOLERANCE AT MMC, MAX DEVIATION

EXAMPLE 2

Ø2.000 +.005 −.000

-A-

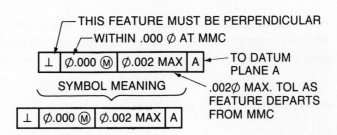

THIS FEATURE MUST BE PERPENDICULAR
WITHIN .000 ∅ AT MMC

⊥ | ∅.000 Ⓜ | ∅.002 MAX | A

TO DATUM PLANE A

.002∅ MAX. TOL AS FEATURE DEPARTS FROM MMC

SYMBOL MEANING

⊥ | ∅.000 Ⓜ | ∅.002 MAX | A

MEANING

DATUM PLANE →

POSSIBLE DIRECTION OF THE FEATURE AXIS

ACTUAL FEATURE SIZE	PERPENDICULARITY TOLERANCE DIAMETER ∅ ALLOWED
2.000 MMC	.000
2.001	.001
2.002	.002
2.003	.002
2.004	.002
2.005 LMC	.002

WHEN THE FEATURE IS AT MMC (2.000) ITS AXIS MUST BE PERPENDICULAR TO THE DATUM PLANE. AS THE FEATURE SIZE DEPARTS FROM MMC (GETS LARGER) AN INCREASE IN TOL IS PERMITTED EQUAL TO THE AMOUNT OF THAT DEPARTURE. UP TO .002 MAX.

ANGULARITY

Definition. Angularity is the condition of a surface, axis, or median plane which is at the specified angle (other than 90°) from a datum plane or axis.

ANGULARITY TOLERANCE

Angularity tolerance is the distance between two parallel planes, inclined at the specified angle to a datum plane or axis, within which the toleranced surface, axis, or median plane must lie.

ANGULARITY APPLICATION

The example shows a part with a surface angular requirement. Note that the symbol is interpreted as "This surface must be at 45° angle in relation to datum plane A within .005 wide tolerance zone in relation to datum plane A."

The interpretation shows how the tolerance zone is established. Note that the angular tolerance zone is at 45° BASIC (exact) from the datum plane A. To be acceptable, the entire angular surface must fall within this tolerance zone. The angular surface must be contained within the limits of part size.

Note in the lower portion of the interpretation that the tolerance zone is also affected and established by the surface itself. That is, the surface extremities actually determine one plane of the tolerance zone as it bottoms out while the plane is inclined at an exact 45° angle with reference to the datum plane.

PLANE SURFACES

EXAMPLE

MEANING

NOTE* Part must be within *size* tolerance limits. The angular tolerance zone, composed of two parallel planes .005 apart, is 45° BASIC to the datum plane A. This tolerance zone is established by contact of the outermost of the two planes with the extremities of the angular surface and with the other plane parallel and inward at .005 distance. The entire surface must fall within this tolerance zone to be acceptable. Actually, this also controls the *flatness* of the surface to .005. Note that the angular surface extremities must be within both size and angular tolerance. See the examples below.

Where angularity of a feature with *size* is to be specified, the median plane or axis of the feature becomes the basis for control. In this case, an *orientation* tolerance characteristic is used as a positional tolerance control.

Figure 1 illustrates an RFS application of a slot in an angular relationship to three datums. The produced slot median or center plane must be within the .005 total wide tolerance zone regardless of the actual size of the slot or datum. The tolerance zone extremities are defined as two parallel planes parallel to the median plane and perpendicular to datum A, with the vertex of the median plane angle established by the centers of datums B and C, RFS.

In this instance, the orientation characteristic, *angularity,* is used in a manner which can be considered a positional relationship. Therefore the position characteristic should have been used. When position is used, the datums and the tolerance require Ⓢ modifiers to state the requirement as RFS, e.g.,

In our example, the RFS angularity specification appears most direct and simple although the multiple datum references complicate the use of the angularity characteristic. Position and its conventional relationship to a datum reference frame is considered more correct in this case.

Figure 2 illustrates an MMC application of a slot in an angular relationship to three datums. The produced slot median plane must lie within the .005 total wide tolerance zone when the slot and datums B and C are at MMC size. The tolerance zone extremities are defined as two parallel planes parallel to the median plane and perpendicular to datum A, with the vertex of the median plane angle established by the centers of datums B and C at MMC size. As the inside slot departs from MMC size (gets larger), an amount equal to that departure is added to the total wide tolerance zone of the slot. At the slot high limit the total wide tolerance zone is .011. As the datums B and C depart from MMC size, further tolerance is added to the total tolerance zone of the slot. Due to the geometry of this part, the amount of added tolerance available relative to the datum departures from MMC size may be difficult to determine. Use of a functional gage will automatically provide the permissible amount of additional tolerance based on actually produced feature size. In this instance the orientation characteristic, angularity, is again used in a manner which can be considered a positional relationship. Therefore the position characteristic should have been used.

Figure 3 illustrates an RFS application of a hole in an angular relationship with three datums. The produced hole axis must lie within the .005 total wide tolerance zone regardless of the actual size of the hole or the datums. The tolerance zone extremities are defined as two parallel planes parallel to the axis and perpendicular to datum A, with the vertex of the axis angle established by the centers of datums B and C, RFS.

It should be noted that this type of angularity tolerancing controls the feature axis only in angular orientation of the view shown. Position control relating the feature to appropriate datums would provide a more suitable control of a feature of this type.

MEDIAN PLANE

EXAMPLE (RFS)

MEANING

FIGURE 1

.250 ±.003

| ∠ | .005 | A | B | C |

.0025

.005 TOTAL WIDE TOL ZONE, RFS ⊥ TO A

45°

45° BSC

CENTER OF B, RFS

CENTER OF C, RFS

.250 ±.002

-C-

⌀1.500 ±.003

-B-

-A-

EXAMPLE (MMC)

MEANING

FIGURE 2

.250 ±.003

| ∠ | .005 Ⓜ | A | B Ⓜ | C Ⓜ |

BELOW PREFERRED (SEE EXPLANATION IN TEXT)

| ⊕ | .005 Ⓜ | A | B Ⓜ | C Ⓜ |

.0025 (AT MMC) .0055 AT HIGH LIMIT

.005 TOTAL WIDE TOL ZONE AT MMC, .011 TOTAL WIDE TOL ZONE AT HIGH LIMIT SIZE OF SLOT * ⊥ TO A

45°

45° BSC

CENTER OF B AT MMC

CENTER OF C AT MMC

.250 ±.002

-C-

⌀1.500 ±.003

-B-

-A-

*SLOT TOL ZONE HAS FURTHER INCREASE RELATIVE TO DEPARTURE FROM MMC SIZE OF DATUMS B & C.

AXIS

EXAMPLE (RFS)

MEANING

FIGURE 3

⌀.250 ±.003

| ∠ | .005 | A | B | C |

.0025

.005 TOTAL WIDE TOL ZONE, RFS ⊥ TO A

45°

45° BSC

CENTER OF C, RFS

.250 ±.002

-C-

⌀1.500 ±.003

-B-

-A-

CENTER OF B, RFS

PARALLELISM //

Definition. Parallelism is the condition of a surface or axis which is equidistant at all points from a datum plane or axis.

PARALLELISM TOLERANCE

A parallelism tolerance specifies:

1.   a tolerance zone defined by two planes or lines parallel to a datum plane (or axis) within which the considered feature (axis or surface) must lie (see Figs. 1 and 2);

2.   a cylindrical tolerance zone parallel to a datum axis within which the axis of the feature under consideration must lie (see Fig. 3).

FIGURE 1   FIGURE 2   FIGURE 3

PARALLELISM APPLICATION

Note in the following example that the bottom surface has been selected as the datum and the top surface is to be parallel to datum plane A within .002.

The Meaning beneath the example clarifies the symbol: It reads, "This feature must be parallel within .002 to datum plane A."

The lower example illustrates the tolerance zone and the manner in which the surface must fall within the tolerance zone to be acceptable. Note that the tolerance zone is established parallel to the datum plane A. Note also that the parallelism tolerance, when applied to a plane surface, controls flatness if a flatness tolerance is not specified (that is, the implied flatness will be *at least* as good as the parallelism).

SURFACE TO DATUM PLANE

EXAMPLE

SYMBOL MEANING

// .002 A ← TO DATUM PLANE A

WITHIN .002 WIDE TOL ZONE

THIS SURFACE MUST BE PARALLEL

MEANING

.002 WIDE TOL ZONE PARALLEL TO DATUM A

.XXX

DATUM A

THE SURFACE MUST BE WITHIN THE SPECIFIED TOLERANCE OF SIZE AND MUST LIE BETWEEN TWO PLANES .002 APART WHICH ARE PARALLEL TO THE DATUM PLANE.

EXAMPLE

// .005 A

.300 ±.010

-A-

MEANING

.005 TOL ZONE
↑.310
PART AT HIGH SIZE LIMIT

.005 TOL ZONE
↑.300
PART AT NOMINAL SIZE

.005 TOL ZONE
↑.290
PART AT LOW SIZE LIMIT

(POSSIBLE VARIATION OF PARALLELISM TOLERANCE ZONE WITHIN SIZE TOLERANCE RANGE)

PARALLELISM—FEATURES OF SIZE

This page and the one following cover the application of parallelism tolerancing to features of size. Note how the MMC applications give the possibility of greater tolerance.

FEATURE AT RFS, DATUM A PLANE

EXAMPLE

SYMBOL MEANING

// .005 A ← TO DATUM PLANE A

WITHIN .005 WIDE TOL ZONE, RFS

THIS FEATURE MUST BE PARALLEL

MEANING

.005 TOL ZONE.

THE FEATURE AXIS MUST BE WITHIN THE SPECIFIED TOLERANCE OF LOCATION. REGARDLESS OF THE ACTUAL SIZE OF THE FEATURE, ITS AXIS MUST LIE BETWEEN TWO PARALLEL PLANES .005 APART WHICH ARE PARALLEL TO THE DATUM PLANE A.

FEATURE AT MMC, DATUM A PLANE

EXAMPLE

Ø.250 +.003 −.000

// .005 Ⓜ A

SYMBOL MEANING

// .005 Ⓜ A ← TO DATUM PLANE A

WITHIN .005 WIDE TOL ZONE AT MMC

THIS FEATURE MUST BE PARALLEL

MEANING

WIDE TOL ZONE

ALLOWABLE DEVIATION

ACTUAL FEATURE SIZE	PARALLELISM TOL WIDTH ALLOWED
.250 (MMC)	.005
.2505	.0055
.251	.006
.2515	.0065
.252	.007
.2525	.0075
.253 (LMC)	.008

FEATURE AXIS MUST BE WITHIN SPECIFIED TOL OF LOCATION & HOLE (AT MMC) AXIS MUST LIE BETWEEN TWO PARALLEL PLANES .005 APART WHICH ARE PARALLEL TO THE DATUM PLANE A.

WHEN FEATURE SIZE DEPARTS FROM MMC (GETS LARGER), AN INCREASE IN THE PARALLELISM TOL IS ALLOWED EQUAL TO THE AMOUNT OF THAT DEPARTURE.

FEATURE AT RFS, DATUM FEATURE AT RFS

EXAMPLE

SYMBOL MEANING

```
// | ⌀.005 | A
```
⟶ TO DATUM AXIS A, RFS

WITHIN ⌀.005 TOL ZONE, RFS

THIS FEATURE MUST BE PARALLEL

MEANING

POSSIBLE DIRECTION FOR THIS FEATURE AXIS

.005 DIAMETER PARALLELISM TOL ZONE

PARALLEL

DATUM AXIS

THE FEATURE AXIS MUST BE WITHIN THE SPECIFIED TOLERANCE OF LOCATION. REGARDLESS OF THE ACTUAL SIZE OF THE FEATURE, ITS AXIS MUST LIE WITHIN A .005 DIAMETER (CYLINDRICAL) TOLERANCE ZONE WHICH IS PARALLEL TO THE DATUM AXIS.

FEATURE AT MMC, DATUM FEATURE AT RFS

(Note that the ⌀ symbol is required to indicate a diameter (cylindrical) tolerance zone.)

EXAMPLE

⌀.250 +.003 −.000

```
// | ⌀.005 Ⓜ | A
```

.XX

−A−

SYMBOL MEANING

```
// | ⌀.005 Ⓜ | A
```
⟶ TO DATUM AXIS A, RFS

WITHIN .005⌀ TOL ZONE AT MMC

THIS FEATURE MUST BE PARALLEL

MEANING

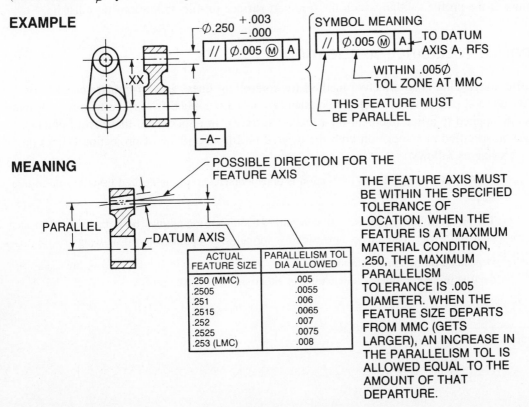

POSSIBLE DIRECTION FOR THE FEATURE AXIS

PARALLEL

DATUM AXIS

ACTUAL FEATURE SIZE	PARALLELISM TOL DIA ALLOWED
.250 (MMC)	.005
.2505	.0055
.251	.006
.2515	.0065
.252	.007
.2525	.0075
.253 (LMC)	.008

THE FEATURE AXIS MUST BE WITHIN THE SPECIFIED TOLERANCE OF LOCATION. WHEN THE FEATURE IS AT MAXIMUM MATERIAL CONDITION, .250, THE MAXIMUM PARALLELISM TOLERANCE IS .005 DIAMETER. WHEN THE FEATURE SIZE DEPARTS FROM MMC (GETS LARGER), AN INCREASE IN THE PARALLELISM TOL IS ALLOWED EQUAL TO THE AMOUNT OF THAT DEPARTURE.

APPLIED TO INDIVIDUAL FEATURES (NO DATUM) OR RELATED FEATURES (USING DATUMS)

Profile tolerancing is of two varieties and involves the characteristics below. According to the design requirement, these characteristics may be applied to an individual feature, such as a *single surface* or *element,* or to related features, such as a *single surface* or *element* relative to a datum or datums.

(PROFILE OF A LINE)   (PROFILE OF A SURFACE)

See following pages for details of application.

PROFILE
METHOD OF SPECIFYING

Definition. Profile tolerancing specifies a uniform boundary along the true profile within which the elements of the surface must lie.

PROFILE TOLERANCE

A profile tolerance (either bilateral or unilateral) specifies a tolerance zone, always* intended and measured normal to the basic profile (applicable to the view in which drawn) at all points of the profile, within which the true part surface profile or line profile must lie.

PROFILE TOLERANCE APPLICATION

Profile tolerancing is an effective method for controlling lines, arcs, irregular surfaces, or other unusual part profiles. Profile tolerances are usually applied to surface features but may also be applied to a line (element on a feature surface). In either case, these requirements must be specified in association with the desired profile in a plane of projection (view) on the drawing as follows:

a.   An APPROPRIATE VIEW or section is drawn which shows the desired basic profile in true shape.

b.   The profile is defined by basic dimensions. This dimensioning may be in the form of located radii and angles, coordinate dimensioning, formulas, or undimensioned drawing.

c.   Depending on design requirements, the tolerance may be divided bilaterally to both sides of the true profile or applied unilaterally to either side of the true profile.

---

*Unless otherwise specified.

Where an equally disposed bilateral tolerance is intended, it is only necessary to show the feature control symbol with a leader directed to the surface. For an unequally disposed or unilateral tolerance, phantom lines are drawn parallel to the true profile to clearly indicate the tolerance zone inside or outside the true profile. One end of a dimension line is extended to the feature control symbol.

BILATERAL   UNILATERAL   UNILATERAL

d.   Other appropriate dimensions, as well as the applicable feature control symbol, are added. The SYMBOL should be applied in a view in which the surface or lines to be controlled is shown as a profile and which pictorially represents the desired feature orientation and relationship.

e.   If some limits on a drawing are expressed by a profile tolerance and others by regular tolerance dimensions, the extent or limitation of the profile tolerance must be clearly indicated by reference letters applied at the extremities of the profile controlled portion and a notation (for example, BETWEEN A & B) added with the profile tolerance symbol or notes. Where a profile tolerance applies *all around* the profile of the part, the symbolic representation (⏚) is specified or the notation ALL AROUND is placed beneath the feature control frame.

Common surface profile tolerancing includes a combination of FORM, ORIENTATION, and SIZE controls, with the profile basic dimensions amounting to overall dimensions as well. In this situation, the standard limits of size interpretation, which require that "form" control must be contained within "size" control (i.e., Rule 1 on page 22), do *not* apply.

Surface profile tolerancing may also be associated with conventional size dimensions and tolerances. The profile itself, however, must be controlled by basic dimensions and the profile tolerance zone. Under this condition the standard limits of size interpretation (Rule 1) do apply and the profile tolerance zone must be contained *within* the *size* tolerance zone. If, for example, the bilateral profile method is used, some portions of the profile tolerance zone may be sacrificed if the controlled feature is at its extreme size limit at that point of the profile.

Since profile tolerancing is a control of *surface form or orientation,* no modifier (MMC) can be used on the feature controlled. Where a datum reference is used, it is usually intended to apply RFS. However, there may be unique applications permitting the use of an MMC datum to expedite gaging provided there is no detrimental effect on the design requirement (e.g., a functional shaft size as related to a cam profile).

GAGING

The example meaning shows the basic gaging principle of measurement *normal to the basic profile.* Gage traverse or part rotation techniques could be used. Where overall size permits, optical comparator (or similar) techniques are usually the most economical and effective method of inspecting this type of part. These techniques make it possible to compare a blown-up (10X, 20X, etc.) projected shadow or profile of the part with an optical gage chart containing accurately scribed profile tolerance zone limits.

Note, however, that shadow image of the part profile which is seen on the comparator screen will be the extreme profile and may not represent the entire surface. Focus adjustment and surface illumination methods (available on some comparators) may be able to account for the entire surface. If necessary, further inspection may be required to account for surface irregularities which are below the extreme profile or not made visible by the method used.

A larger part exceeding the normal limitations of the optical chart size might be checked in segments or with the use of tracer and reticle.

Computerization techniques, either independently applied or in conjunction with optical methods, also provide possible verification options.

TWO TYPES OF PROFILE TOLERANCE

In practice, a profile tolerance may be applied to either an entire surface or to individual line element profiles taken at various cross sections through the part. The two types or methods of controlling profile are:

a.   Profile of a surface ⌓
The tolerance zone established by profile of a surface tolerance is a three-dimensional zone or total control across the entire length and width or circumference of the feature; it may be applied to parts having a constant cross section or to parts having a surface of revolution. Usually profile of a surface requires datum references.

b.   Profile of a line ⌒
The tolerance zone established by profile of a line tolerance is a two-dimensional zone extending along the length of the considered feature; it may apply to the profiles of parts having a varying cross section such as a propeller, aircraft wing, nose cone, or to random cross sections of parts where it is not desired to control the entire surface as a single entity. Profile of a line may, or may not, require datum references.

⌒ PROFILE OF A SURFACE

EXAMPLE

MEANING

---

*.010 total wide zone (.005 each side of true profile). The surface is between the two profile boundaries .010 apart, equally disposed about the true profile, which are perpendicular to datum plane A and positioned with respect to datum planes B and C.

PROFILE
PROFILE OF A SURFACE

Where the profile tolerance is to be used in controlling the entire surface profile, the symbolic representation may be specified or the words "ALL AROUND" should be added below the profile tolerance.

ZONE TOLERANCE AROUND ENTIRE PROFILE

EXAMPLE

GAGING

The example illustrates the basic gaging principle of measurement *normal to the true profile* as oriented from datum A. Where overall size permits, optical "comparator" (or similar) techniques are usually the most economical and effective method of inspecting this type of part (see also GAGING of preceding example).

Where the profile tolerance is to be used in controlling all, or large portions, of the surface profile, specific notations (such as angular degrees on a cam) should be added with the profile tolerance symbol or note.

This type of tolerance control is particularly useful on cams or similar parts.

DIFFERENT ZONE TOLERANCE AROUND ENTIRE PROFILE

EXAMPLE

CAM DEGREES BASIC	CAM RADIUS BASIC
0°	.390
52°	.390
59°	.376
67°	.362
75°	.359
92.5°	.355
145°	.338
180°	.323
213°	.313
269°	.297
299°	.287
336°	.194

MEANING

TOTAL WIDE TOL ZONE .002 FROM 67° TO 299° .004 FROM 0° TO 67° .004 FROM 299° TO 360° (.001 & .002 RESPECTIVELY EACH SIDE OF TRUE PROFILE) WITHIN WHICH ACTUAL SURFACE PROFILE MUST LIE

GAGING

Optical comparator (or similar) techniques are ideal for this part. A transparent chart made to 20X, 50X, etc. size with accurately scribed lines defining the tolerance zones permits a direct surface profile check of the magnified part with the chart. If the surface profile of the part lies within the scribed tolerance zones when oriented to the datum B and C axes and while bottoming on datum A, it is acceptable.

PROFILE
PROFILE OF A LINE

The profile of a surface controls the entire surface, or all the elements of the surface, within a uniform tolerance zone as established from the true profile. However, where line elements of a surface are to be specifically controlled or controlled as a refinement of size or surface profile control, the profile of a line characteristic may be used. Line profile control is applied in a manner similar to the application of surface profile.

The example at the right illustrates line profile control as a refinement of size. As with surface profile, line profile must be shown in the drawing view in which the control applies. The tolerance zone is established in the same way as surface profile. However, its tolerance zone is disposed about each element of the surface. Therefore the tolerance zone applies for the full length of each element of the surface (in the view in which it is shown), but only for the width or height at a cutting plane bisecting the element. This cutting plane must be considered to be perpendicular or parallel to the part orientation within its size tolerances. Where datum references are specified, the orientation is with respect to the datums to give a more specific relationship. Each actual line element of the controlled profile may vary within its prescribed tolerance zone, but the element-to-element control may vary within the entire size tolerance.

Line profile control is illustrated in the example. Accuracy and consistency of the individual elements at the part surface are critical, but variation from element to element is less critical. More lenient control within the size tolerances is adequate in the direction perpendicular to the line profile. The requirement to hold an accurate line profile on an irregularly shaped part within a more lenient size control is similar to holding straightness within a more lenient size control on a square, rectangular, or cylindrical part.

Surface profile and line profile may be applied to the same feature, where applicable—for example, in situations in which the line elements are to be controlled more closely than the surface as a whole. Combined surface and line profile control may be used and specified as shown in the figure below.

EXAMPLE

SYMBOL MEANING

THE ACTUAL *SURFACE* MUST LIE WITHIN THE SPECIFIED TOLERANCE ZONE OF .010 WITH RESPECT TO DATUM A

ANY *LINE* ELEMENT OF THE SURFACE MUST LIE WITHIN THE SPECIFIED TOLERANCE ZONE OF .005 WHICH IS CONTAINED WITHIN, AND OF THE SAME GEOMETRIC FORM, AS THE .010 TOLERANCE ZONE

EXAMPLE

MEANING

Any profile line element of the surface (elements in the view in which the symbol is shown and element represented by the true profile) must be within the specified .006 tolerance zone which is unilaterally disposed from the true profile. The *profile* tolerance zone must be contained within the *size* tolerance zone.

GAGING

The illustration above shows the basic gaging principle of measurement *normal to the basic profile* and in an imaginary cutting plane which bisects the line element while oriented basically perpendicular and parallel within the size tolerances.

Dial indicator, master template, and optical comparator (or similar) techniques can be used (see also GAGING in preceding examples).

COPLANAR SURFACES

Profile tolerancing may be used to specify coplanarity of two or more surfaces where it is desired to treat these surfaces as a single interrupted or noncontinuous surface. In such an application, a control is provided similar to that achieved by a flatness tolerance applied to a single plane surface. Coplanarity is the condition of two or more surfaces having all elements in one plane.

As shown in Example 1, the profile of a surface tolerance establishes a tolerance zone defined by two parallel planes within which the considered surfaces must lie. No datum reference is stated (as in the case of flatness) since the orientation of the tolerance zone is established from the contact of the part against a reference standard; the datum is established by the considered surfaces themselves.

Where more than two surfaces are involved, it may be desirable to identify which specific surface (or surfaces) is to be used as a datum and to establish the tolerance zone. Where necessary, datum target methods could also be used. Where a datum (or datums) is used, it is understood that the tolerance zone established applies to all coplanar surfaces, including the datum surfaces, unless otherwise specified as shown in Example 2.

OFFSET SURFACES

As a variation of coplanar surfaces, the principles described above may be extended to offset surfaces as shown in the example. In this instance, the variation from coplanar is a desired amount (.105) and is stated by a basic dimension.

COPLANAR SURFACES

EXAMPLE 1

2 SURFACES

MEANING

.003 WIDE
TOL ZONE

THE SURFACES MUST BE WITHIN
THE SPECIFIED TOL OF SIZE AND
MUST BOTH LIE BETWEEN TWO
PARALLEL PLANES .003 APART.

EXAMPLE 2

MEANING

.003 WIDE
TOL ZONE

DATUM PLANE A

THE SURFACE MUST BE WITHIN THE
SPECIFIED TOL OF SIZE AND MUST
LIE BETWEEN TWO PARALLEL
PLANES .003 APART AS
ESTABLISHED RELATIVE TO DATUM
PLANE A

OFFSET SURFACES

EXAMPLE

.105

MEANING

.0015

DATUM PLANE A   .105   (PARALLEL
TO A)

.003 WIDE
TOL ZONE

THE SURFACE MUST BE WITHIN THE
SPECIFIED TOL OF SIZE AND MUST
LIE BETWEEN TWO PARALLEL
PLANES .003 APART AS
ESTABLISHED RELATIVE TO
DATUM A

CONICITY

Profile of a surface tolerance can be applied to conical parts to control conicity. It may be applied as an independent control of form (Example 1, next page) or as a combined control of form and orientation with respect to a datum feature and datum axis (Example 2, next page). It could also be used, when necessary, as profile ALL AROUND control, which can control size and form simultaneously using basic dimensions to define the conical shape.

In Example 1, profile of a surface is used as a *form* control (of the 20° cone) as a refinement within the size control. This is comparable to a cylindricity tolerance controlling *form* of a cylinder as a refinement within its size control. *No* datum is required nor proper in this case.

In Example 2, profile of a surface is used as a *form and orientation* control (of the 15° cone) as a refinement within the size control and relative to datum A. The added notation ALL AROUND could be used if desired for clarification but is not necessary here. In this case, profile tolerance is applied to a surface of revolution as a design preference (hypothetically) over other methods (i.e., runout).

See DATUM section for more information on datums and datum precedence (primary, secondary, tertiary datums).

CONICAL SURFACES

EXAMPLE 1

MEANING

EXAMPLE 2

MEANING

RUNOUT TOLERANCES—RELATED FEATURES USING DATUMS

A runout tolerance is a relationship between surfaces or features, therefore, a datum (or datums) is required.

Runout tolerance may be applied in two different ways using the characteristics shown below.

(CIRCULAR RUNOUT)   (TOTAL RUNOUT)

See following pages for details of application.

RUNOUT (CIRCULAR AND TOTAL)

Definition. Runout is the composite deviation from the desired form and orientation of a part surface of revolution during full rotation (360°) of the part on a datum axis.

RUNOUT TOLERANCE

Runout tolerance states how far an actual surface or feature is permitted to deviate from the desired form and orientation implied by the drawing during full rotation (360°) of the part on a datum axis.

RUNOUT APPLICATION

Runout tolerancing is a method used to control the composite surface effect of one or more features of a part relative to a datum axis. Runout tolerance is applicable to rotating parts in which this composite surface control is based on the part function and design requirement. A runout tolerance always applies on an RFS basis; namely, size variation has no effect upon the runout tolerance compliance.

Each considered feature must be within its individual runout tolerance when rotated 360° about the datum axis. The tolerance specified for a controlled surface is the total tolerance or full indicator movement (FIM) in terms of common inspection criteria. Former terms, full indicator reading (FIR) and total indicator reading (TIR), have the same meaning as FIM.

As is seen at right, the basis of runout tolerance control is the datum axis of the part. Surfaces controlled may be those constructed *around* a datum axis, or those constructed at *right angles* to a datum axis. As is also seen, the datum axis is established from a datum feature.

BASIS OF CONTROL OF RUNOUT TOLERANCE

SURFACES AT RIGHT
ANGLES TO A DATUM AXIS

DATUM FEATURE

DATUM AXIS
(ESTABLISHED FROM
DATUM FEATURE)

SURFACES CONSTRUCTED
AROUND A DATUM AXIS

TYPES OF FEATURES APPLICABLE TO A RUNOUT TOLERANCE

↗ CIRCULAR RUNOUT

EXAMPLE

Ø.XXX ±.XXX

↗ | .002 | A

Ø.XXX ±.XXX

Ø.XXX ±.XXX

−A−

SYMBOL MEANING

↗ | .002 | A — IN RELATION
TO DATUM AXIS A

WITHIN .002 WIDE
TOL ZONE (FIM)

EACH CIRCULAR
ELEMENT OF THE
FEATURE MUST BE
WITHIN THE RUNOUT
TOL

MEANING

FIM .002

.002 FIM — EACH CIRCULAR
ELEMENT
INDIVIDUALLY

BEARING OR
COLLET

DATUM AXIS A

DATUM FEATURE A

SIMULATED DATUM A
CYLINDER

ROTATE PART

⟋⟋ TOTAL RUNOUT

EXAMPLE

MEANING

The datum axis may be established by a single diameter (cylinder) of sufficient length, two diameters with sufficient axial separation, or a diameter and a face surface which is at right angles to it. Features selected as datums should, as much as possible, be functional to the part requirement (e.g., bearing mounting diameters, etc.).

TYPES OF RUNOUT CONTROL

The two types of runout control are *circular* runout and *total* runout. Selection of the proper type is based upon the design requirement and manufacturing considerations. The fundamental difference between the two types are illustrated on p. 81 and above with detailed examples following in this section. Note that with either method, the collective or composite control of various form and orientation variations of the part provides a more direct representation of part functions, integrates manufacturing operations, and minimizes inspection setup requirements.

COAXIAL FEATURES—SELECTION OF PROPER CONTROL

There are three methods of controlling interrelated coaxial features:

1. RUNOUT TOLERANCE (circular or total)   (RFS)

2. POSITION TOLERANCE   (MMC)

3. CONCENTRICITY TOLERANCE   (RFS)

Any of the above methods provides effective control. However it is important to select the *most appropriate* one to both meet the design requirements and provide the most economical manufacturing conditions. (See also details of preceding sections).

Below are recommendations to assist in selecting the proper control:

If the need is to control only CIRCULAR cross-sectional elements in a composite relationship to the datum axis, RFS, e.g., multidiameters on a shaft, use

CIRCULAR RUNOUT ↗   **EXAMPLE**   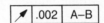

(This method controls any composite error effect of circularity, concentricity, and circular cross-sectional profile variations.)

If the need is to control the TOTAL cylindrical or profile surface in composite relative to the datum axis' RFS, e.g., multi-diameters on a shaft, bearing mounting diameters, etc., use

TOTAL RUNOUT ↗↗   **EXAMPLE**

(This method controls any composite error effect of circularity, cylindricity, straightness, coaxiality, angularity, and parallelism.)

NOTE Runout is always implied as an RFS application. It cannot be applied on an MMC basis, since an MMC situation involves functional interchangeability or assemblability (probably of mating parts), in which case POSITION tolerance would be used. See below.

If the need is to control the total cylindrical or profile surface and its axis in composite location relative to the datum axis on an MMC basis, e.g., on mating parts to assure interchangeability or assemblability, use

POSITION ⊕   **EXAMPLE**

If the need is to control the *axis* of one or more features in composite relative to a *datum axis*, RFS, e.g., to control balance of a rotating part, use

CONCENTRICITY ◎   **EXAMPLE** | ◎ | ⌀ | .002 | A–B |

NOTE Concentricity is always implied as an RFS application. Variations in size (departure from MMC size, out-of-circularity, out-of-cylindricity, etc.) do not in themselves conclude *axis* error.

CIRCULAR RUNOUT ↗

PART MOUNTED ON FUNCTIONAL DIAMETER (DATUM)

Circular runout provides a composite control of circular elements of a surface. Circular runout is normally a less complex requirement than total runout. The tolerance is applied independently at any circular cross section or measuring position on the part as it rotates through 360°. Circular runout should be considered when the part function and manufacturing requirements are satisfied by this type of control. Where more complete control of *all* elements in composite is necessary, total runout should be considered.

Circular runout controls composite variations of circularity and cross-sectional form variations of the surface at each circular element where applied to surfaces constructed around a datum axis. It controls circular elements of the surface (wobble) where applied to surfaces constructed at right angles (perpendicular) to a datum axis.

The examples at right utilize circular runout control. The upper example applies circular runout to the angular surface controlling the individual circular elements of the part as it rotates.

When circular runout is to be applied at specified locations, it must be so stated on the drawing. The lower example at right illustrates this application.

Please see following pages and examples in this section for further applications of and considerations on circular runout and for circular runout and total runout used on the same part.

EXAMPLE

SYMBOL MEANING

IN RELATION TO DATUM AXIS A

WITHIN .001 WIDE TOL ZONE (FIM)

EACH CIRCULAR ELEMENT MUST BE WITHIN THE RUNOUT TOLERANCE

MEANING

FIXED POSITION OF INDICATOR FOR EACH ROTATION (360°)

.001 FIM MAX

ANY GIVEN RADIUS

SINGLE CIRCULAR ELEMENT

THE FEATURE MUST BE WITHIN THE SPECIFIED TOLERANCE OF SIZE AT ANY MEASURING POSITION, EACH CIRCULAR ELEMENT OF THE SURFACE MUST BE WITHIN .001 FULL INDICATOR MOVEMENT WHEN THE PART IS ROTATED ONE FULL ROTATION ABOUT THE SPECIFIED DATUM AXIS WITH THE INDICATOR FIXED IN A POSITION NORMAL TO THE SURFACE. (THIS DOES NOT CONTROL FORM OF THE TOTAL SPECIFIED SURFACE AREA, BUT ONLY CONTROLS THE RUNOUT OF EACH CIRCULAR ELEMENT.)

EXAMPLE

APPLIES AT Ø.50

SYMBOL MEANING

IN RELATION TO DATUM AXIS A

WITHIN .001 WIDE TOL ZONE (FIM)

THE CIRCULAR ELEMENT AT Ø.50 MUST BE WITHIN THE RUNOUT TOLERANCE

MEANING

FIXED POSITION OF INDICATOR PARALLEL TO THE AXIS FOR 360° ROTATION

Ø.50

.001 FIM

SINGLE CIRCULAR ELEMENT

THE CIRCULAR ELEMENT OF THE SURFACE AT Ø.50 MUST BE WITHIN .001 FULL INDICATOR MOVEMENT WHEN THE PART IS ROTATED ONE FULL ROTATION ABOUT THE SPECIFIED DATUM AXIS WITH THE INDICATOR FIXED IN A POSITION PARALLEL TO THE AXIS. (THIS DOES NOT CONTROL PERPENDICULARITY, BUT CONTROLS ONLY THE LATERAL RUNOUT [WOBBLE] OF EACH CIRCULAR ELEMENT AT THE SPECIFIED SURFACE LOCATION.)

PART MOUNTED ON TWO FUNCTIONAL DIAMETERS (DATUMS)

Where multiple diameters of a rotating part are to be controlled relative to a datum axis, the datum axis can be established by the features which will provide the functional mounting of the part at assembly. These are called multiple datums.

Two diameters (cylinders) are used on the part at right to establish the datum axis of relationship. These diameters could represent the bearing mounting features and thus they establish the datum axis of rotation. Note that two datum features are selected (C and D). By stating the two datums simultaneously with a dash line between them, the relationship of all the other features so designated relate to their common datum axis C–D. The precedence of datums C and D are equal in this case.

Any two surfaces given runout tolerances about the datum axis are to be individually within their stated runout tolerance; collectively they are related to each other within the sum of their indicator readings.

A runout tolerance specified for the datum feature has no effect on the considered features related to it.

PART MOUNTED ON TWO FUNCTIONAL DIAMETERS (DATUMS)

EXAMPLE

(SIZE DIMENSIONS & TOLERANCES NOT SHOWN FOR SIMPLICITY)

MEANING

WHEN MOUNTED ON DATUMS C AND D,
DESIGNATED SURFACES MUST BE WITHIN CIRCULAR
RUNOUT (⚹) TOLERANCE SPECIFIED, RFS

CIRCULAR RUNOUT ↗ AND TOTAL RUNOUT ↗↗

PART MOUNTED ON TWO FUNCTIONAL DIAMETERS (DATUMS) INCLUDING RUNOUT TOLERANCE ON DATUMS

The part shown at right extends the principles of the previous example. It is a shaft of multi-diameters about a common datum axis C–D with each feature, *including the datum features,* stating an individual runout tolerance. It utilizes both circular runout and total runout control.

Total runout controls composite surface variations of circularity, cylindricity, parallelism, straightness, angularity, taper, and profile of a surface where applied to features constructed around a datum axis. Total runout applied to surfaces construced perpendicular to a datum axis controls composite variations of perpendicularity and flatness. The tolerance is applied *simultaneously* at all circular and profile measuring positions as the part is rotated through 360°.

Due to the design requirements involved in this example, certain diameters (the datums) must be given total runout control, whereas the remaining diameters and face surfaces may be controlled with circular runout.

Since in the end assembly of this part, (mount to bearings), the relationship to both datums simultaneously is the result desired, all the features including each datum (C and D) must meet their individual runout tolerances. Note that the runout tolerance of each datum (C and D) individually is relative to their common axis C–D.

Multiple leaders may be used in controlling two or more features with a common runout tolerance as is shown. Runout tolerance may be specified individually or in groups, as is convenient, without affecting the runout tolerance.

PART MOUNTED ON TWO FUNCTIONAL DIAMETERS

EXAMPLE

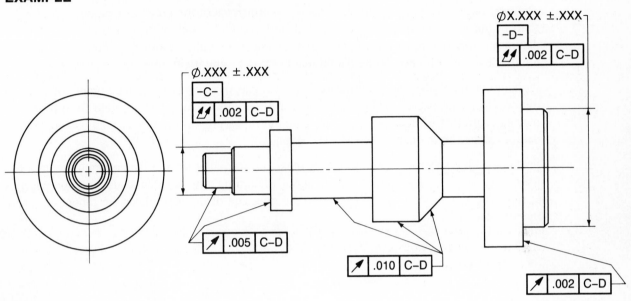

MEANING

ALL READINGS FIM

DATUM FEATURE C
SIM. DATUM C
(CYLINDER)

DATUM AXIS C–D

ROTATE PART

DATUM FEATURE D

DATUM D SIM. (CYLINDER)

—EACH CIRCULAR ELEMENT INDIVIDUALLY—
—DATUM DIAMETERS C AND D ALL ELEMENTS TOGETHER (TOTAL)—

PART MOUNTED ON TWO FUNCTIONAL DIAMETERS (DATUMS) INCLUDING RUNOUT AND CYLINDRICITY TOLERANCE ON DATUM

The part shown at the right further extends the principles of the previous examples. It is a shaft of multi-diameters about a common datum axis C–D, with each feature including the datum features stating an individual runout tolerance. In addition to having both circular and total runout control, each of the datum features has a cylindricity tolerance.

It should be noted that the datum features of this part have controls of three orders of magnitude: size, total runout, and cylindricity (form). In this example, an accurate fit to bearings as controlled by size and form was required in addition to the runout relationship of the datum features to the datum axis (C–D) of rotation.

Features with a specific relationship to another feature rather than to the common datum axis may be so indicated by addition of appropriate datum references (such as datum E on page at right). Note that the inside diameter on the right end of the part relates the runout requirement to datum E.

PART MOUNTED ON TWO FUNCTIONAL DIAMETERS (DATUMS)

EXAMPLE

(SIZE DIMENSIONS & TOLERANCES NOT SHOWN FOR SIMPLICITY)

MEANING

ALL VALUES FIM (TIR, FIR)

WHEN MOUNTED ON DATUMS C AND D, DESIGNATED SURFACES MUST BE WITHIN CIRCULAR RUNOUT (⌁) TOLERANCE SPECIFIED. DATUMS C AND D MUST ALSO BE WITHIN TOTAL RUNOUT (⌁⌁) TOLERANCE SPECIFIED AND CYLINDRICAL WITHIN .0002.

WHEN MOUNTED ON DATUM E, DESIGNATED SURFACE MUST BE WITHIN CIRCULAR RUNOUT (⌁) TOLERANCE SPECIFIED.

PART MOUNTED ON FUNCTIONAL FACE SURFACE (DATUM) AND DIAMETER (DATUM)

Runout tolerancing may be applied to features of rotation where the feature datum references are an axis and a face surface perpendicular to the axis. In such an application, datum precedence is usually considered necessary. The influence of the appropriate feature as a primary datum is determined relative to the design requirement.

In the example at the right, the face surface perpendicular to the axis of rotation is considered the functionally important surface (bottoms on the inner race of bearing) and is thus selected as the primary datum A. Secondary datum diameter (cylinder) B is in contact with the extremities of datum feature B. With the part in this orientation and located functionally on its datums, all related features must meet their individual runout tolerances when the part is rotated about datum axis B.

Note that the primary datum A and secondary datum B features are indicated with that precedence by separate enclosures in the feature control frame (reading left to right).

PART MOUNTED ON FUNCTIONAL FACE SURFACE (DATUM) AND DIAMETER (DATUM)

EXAMPLE

MEANING

.002 WIDE TOL ZONE AT EACH CIRCULAR ELEMENT, FIM

DATUM AXIS B

CIRCULAR RUNOUT ↗ AND FLATNESS ▱

PART MOUNTED ON FUNCTIONAL FACE SURFACE (DATUM) AND DIAMETER (DATUM)

The part shown at right extends the principles of runout tolerancing. It illustrates a situation in which the large flat mounting surface requires flatness control to assure desired accuracy of the primary datum surface as it mounts and orients the remainder of the part to proper functional relationships. In this instance it can be assumed that this method was most representative of part end function.

The remainder of the part relates to the datum system C and D, with C being the primary datum for attitude of the part, and D the secondary datum for establishing axis of rotation. Note again that the other features *including datum D* relate to the datum system C and D.

PART MOUNTED ON LARGE FLAT SURFACE (DATUM) AND DIAMETER (DATUM)

EXAMPLE

MEANING

ROTATE PART

WHEN MOUNTED ON DATUMS C AND D, DESIGNATED SURFACES MUST BE WITHIN CIRCULAR RUNOUT (⚹) TOLERANCE SPECIFIED.

SECONDARY DATUM CYLINDER "D" PERPENDICULAR TO DATUM "C"

-D-

.001 ERROR MAX (FIM)

FLAT WITHIN .001 TOTAL

"D" AXIS

PRIMARY DATUM PLANE "C"

-C-

TOTAL RUNOUT
RUNOUT APPLICATION AND ANALYSIS

Typical total runout applications, inspection analysis, and meaning are discussed below and in the following text. An understanding of the fundamentals involved in these examples will provide the reader with the basis for total runout characteristic and use. The methods and principles shown may be readily adapted to circular runout as well, except for the substitution, where appropriate in the text and illustrations, of the circular elements (only) principles. In checking total runout, the datum feature, or features, is normally centered or positioned in an appropriate inspection device (e.g., between centers, collet, chuck, mandrel, vee block, or other centering device) to establish the datum axis from which the total runout relationships are to occur.

PART MOUNTED ON MACHINING CENTERS

EXAMPLE

MEANING

.001 FIM
.0005 FIM
.002 FIM
.0015 FIM
ROTATE PART

SIM. DATUM A
DATUM FEATURE A
DATUM B SIM.
DATUM FEATURE B
DATUM AXIS A–B

PART MOUNTED ON TWO FUNCTIONAL DIAMETERS

EXAMPLE

MEANING 1

TOTAL SURFACE RUNOUT (WOBBLE)

Note that when flat or face surface runout (wobble) is part of the specification as in the case of the example on the preceding page, the runout analysis must consider the need for restricting end movement of the part while checking it. If collets or other holding devices are applied to establish the datum axis, end movement of the part is controlled. Also, specified secondary or tertiary datums (specified faces, shoulders, etc.) may indicate the end location or "stop" surface. However, in the absence of the above conditions and when the flat or face surface runout is to be checked, one should, if possible, "stop" on the surface being checked (see MEANING 1 on the preceding page and continued below), in order to avoid adding the error of *another* surface to the reading. When this is done, however, dependent on the location or placement of the stop, the dial indicator runout reading may need to be *halved* (divided by 2) before comparison with the stated part drawing runout tolerance to determine whether the requirement has been met. Referring to the illustration below, note that to use the fixed factor 2 the stop should be as close as possible to the O.D. of the surface being checked. As the placement of the stop approaches the axis of the mounting datum diameters (see also MEANING 2 and 3), the factor approaches a 1:1 ratio.

"D" = DIRECT DIAL INDICATOR READING WHEN PART ROTATED 180° (TWICE ACTUAL RUNOUT)

$\frac{\text{"D"}}{2}$ = ACTUAL TOTAL SURFACE RUNOUT

PART ROTATED 180°

A–B AXIS (MOUNTED ON A & B)

DATUM A

DATUM B

STOP

NOTE If stop could be placed on mounted dia axis, runout indicator reading would be direct, 1:1. At any location between, would vary from 2:1 (at O.D.) to 1:1 (at axis).

MEANING 2

As an alternative method a full end stop may be used. The end stop must, however, be extremely accurate at 90°(exact within gage tolerances) to the axis of the mounting diameters. See below.

In this case the dial indicator reading taken is direct and can be compared immediately with the stated part drawing runout (wobble) tolerance.

Note also that no error of the part surface or face which is used for "stopping" is introduced into the indicator reading, since the high point or extremity of this surface remains in consistent contact with the full stop gage as the part is rotated.

MEANING 3

If the part has a *closed end* and a stop can be placed at the *axis* of the mounting datum diameters (see illustration below), the runout requirement may be checked without regard for extreme accuracy of the stop. The dial indicator reading taken is direct and can be compared immediately with the stated part drawing runout (wobble) tolerance.

RUNOUT
PARTIAL SURFACE*

Where it is desirable to indicate that a runout tolerance is applicable only at a specific place or portion of a surface feature, it may be shown as indicated in the example below. Basic (exact) dimensions are shown indicating the portion of the surface to which the runout tolerance applies.

A thick chain line is shown slightly off the part with the leader from the feature control frame extended to the concerned surface. The thick chain line means that the stated tolerance applies only at the designated area. The basic dimensions invoke standard inspection accuracy in verifying the requirements. Otherwise, standard interpretation is given to the total runout requirement shown (see previous pages).

EXAMPLE

MEANING

---

*This principle may be applied to other geometric controls as well.

TOLERANCES OF LOCATION

Tolerances of location state the permissible variation in the specified location of a feature in relation to some other feature or datum. Tolerances of location refer to the geometric characteristics: position and concentricity.

In the course of the discussion on location tolerancing, more detail on maximum material condition, datums, basic dimensioning, and the interrelationship of location and orientation tolerancing will be introduced.

Location tolerances involve features of size and relationships to center planes and axes. At least two features, one of which is a size feature, are required before location tolerancing is valid. Where function or interchangeability of mating part features is involved, the MMC principle may be introduced to great advantage. Perhaps the most widely used and best example of the application of this principle is position tolerancing.

The use of the position concept in conjunction with the maximum material condition concept provides some of the major advantages of the geometric tolerancing system.

NOTE Symmetrical features previously controlled by symmetry (≡) now controlled by position (⊕).

POSITION ⊕

Definition. Position is a term used to describe the perfect (exact) location of a point, line, or plane of a feature in relationship with a datum reference or other feature.

POSITION TOLERANCE

A position tolerance is the total permissible variation in the location of a feature about its exact true position. For cylindrical features (holes and bosses) the position tolerance is the *diameter (cylinder)* of the tolerance zone within which the axis of the feature must lie, the center of the tolerance zone being at the exact true position. For other features (slots, tabs, etc.) the position tolerance is the total width of the tolerance zone within which the center plane of the feature must lie, the center plane of the zone being at the exact true position.

POSITION THEORY

We shall now examine the position theory as typically applied to a part for purposes of function or interchangeability. As a means of describing this theory we shall first compare the position system with the bilateral or coordinate system.

Imagine a part with four holes in a pattern which must line up with a mating part to accept screws, pins, rivets, etc. to accomplish assembly, or four holes in a pattern to accept the pins, dowels, or studs of a mating part to accomplish assembly.

The top figure at the right shows the part with a hole pattern dimensioned and toleranced using a coordinate system. The bottom figure shows the same part dimensioned using the position system. Comparing the two approaches, we find the following differences:

1. The derived tolerance zones for the hole centers are square in the coordinate system and round in the position system.

2. The hole center location tolerance in the top figure is part of the coordinates (the 2.000 and 1.750 dimensions). In the bottom figure, however, the location tolerance is associated with the hole size dimension and is shown in the feature control frame at the right. The 2.000 and 1.750 coordinates are retained in the position application, but are stated as BASIC or exact values.

For this comparison, the .005 square coordinate tolerance zone has been converted to an equivalent .007 position tolerance zone. The two tolerance zones are superimposed on each other in the enlarged detail.

The black dots represent possible inspected centers of this hole on eight separate piece parts. We see that if the coordinate zone is applied, only three of the eight parts are acceptable. However, with the position zone applied, six of the eight parts appear immediately acceptable.

COORDINATE SYSTEM

POSITION SYSTEM

IF HOLE SIZE	POSN. TOL
.247 MMC	⌀.007
.248	.008
.249	.009
.250	.010
.251	.011
.252	.012
.253 LMC	.013

The position diameter shaped zone can be justified by recognizing that the .007 diagonal is unlimited in orientation. Also, a cylindrical hole should normally have a cylindrical tolerance zone.

A closer analysis of the representative black dots and their position with respect to the desired exact location clearly illustrates the fallacies of the coordinate system when applied to a part such as that illustrated.

The dot in the upper left diagonal corner of the square zone and the dot on the left outside the square zone are in reality at nearly the same distance from the desired exact center. However, in terms of the square coordinate zone, the hole on the left is unacceptable by a wide margin, whereas the upper left hole is acceptable.

Note, then, that a hole produced off center under the coordinate system has *greater* tolerance if the shift is on the diagonal and not in the horizontal or vertical direction.

Realizing that the normal function of a hole relates to its mating feature in *any* direction (i.e., a hole vs. a round pin), we see that the square zone restriction seems unreasonable and incorrect. Thus the position tolerance zone, which recognizes and accounts for unlimited orientation of round or cylindrical features as they relate to one another, is more realistic and practical.

In normal applications of position principles, the tolerance is derived, of course, from the design requirement, *not* from converted coordinates. The maximum material sizes of the features (hole and mating component) are used to determine this tolerance.

Thus the .007 position tolerance of the example on the preceding page (see also Fig. 1) would normally be based on the MMC size of the hole (.247). As the hole size deviates from the MMC size, the position of the hole is permitted to shift off its "true position" beyond the original tolerance zone to the extent of that departure. The "bonus tolerance" of .013 illustrates the possible position tolerance should the hole be produced, for example, to its high limit size of .253. The tabulation on the preceding page shows the enlargement of the position tolerance zone as the hole size departs from MMC.

Although we have considered only one hole to this point in the explanation, the same reasoning applies to all the holes in the pattern. Note that position tolerancing is also a noncumulative type of control in which each hole relates to its own true or exact position and no error is accumulated from the other holes in the pattern.

Position tolerancing is usually applied on mating parts in cases where fit, function, and interchangeability are the considerations. It provides greater production tolerances, ensures design requirements, and provides the advantages of functional inspection practices as desired.

Functional gaging techniques, familiar to a large segment of industry through many years of application, are fundamentally based on the MMC position concept. It should be clearly understood, however, that functional gages are not mandatory in fulfilling MMC position requirements.

Functional gages are used and discussed in this text for the dual purpose of explaining the principles involved in position tolerancing and of introducing the functional gage technique as a valuable tool. A functional gage can be considered as a simulated master mating part at its worst condition.

Position, although a locational tolerance, also includes form and orientation tolerance elements in composite. For example, as shown in the illustration on page 103, perpendicularity is invoked as part of the control to the extent of the diameter zone, actually as a "cylindrical" zone, for the depth of the hole. Further, the holes in the pattern are parallel to one another within the position tolerance. Various other elements of form are included as a part of the composite functional control provided by position tolerancing. Use of datum references and the relation of the hole pattern to specific surfaces or other features further ensures that the holes (and their tolerance zones) are related to part function and will be uniformly interpreted. For purposes of simplicity in explaining the theory, only *one* datum feature was specified in this example. A complete specification of this requirement would include *two* additional datums (edges of the part). For a more complete discussion on the datum reference frame, see the DATUM section and numerous examples following in this text.

POSITION SYSTEM

The example below further clarifies the position theory; two of the holes on the part shown in the example are enlarged to illustrate the actual effect of feature *size* variation on the previous positional location of the features.

POSITION THEORY

Figure 1 shows the two .250 ± .003 holes at MMC size (or the low limit of their size tolerance) of .247 and with their centers perfectly located in the .007 diameter position tolerance zone. The drawing illustrates the mating part situation represented by a functional or fixed pin. The gage pins are shown undersize an amount equal to the positional tolerance of .007, i.e., at .240 diameter. This represents the maximum permissible offset of the holes within their stated positional tolerance when the hole is at MMC size of .247.

FIGURE 1

Figure 2 shows the two .250 MMC holes offset in opposite directions to the maximum permissible limits of the .007 position tolerance zone. Note that we illustrate the worst condition: the edges of the holes are tangent to the diameters of the simulated mating part or gage pins. The holes are within tolerance and, as can be seen, would satisfactorily pass the simulated mating part condition as represented by the gage pins.

In Fig. 3 the .250 ± .003 holes have been produced to the opposite, or *high* limit (least material condition) size of .253. It can now be seen that when we retain the same offset and tangency of the holes and mating part of the gage pins as shown in Fig. 2, the produced centers of the holes are allowed to shift *beyond* the original .007 tolerance zone to a resulting .013 diameter tolerance zone still providing an acceptable situation.

FIGURE 2

HOLES OFFSET AT MMC

1.750

.007 DIA POSITION
TOL ZONE

2.000

HOLE AT MMC
(LOW LIMIT) ⌀.247

FIGURE 3

HOLES OFFSET AT
HOLE HIGH LIMIT SIZE
(LEAST MAT'L COND.)

1.750

.013 DIA POSITION
TOL ZONE

2.000

HOLE AT
HIGH LIMIT
(LEAST MAT'L COND.)
⌀.253

POSITION TOL = .007
+ TOTAL HOLE TOL = .006

TOTAL POSSIBLE
POSITIONAL TOL   = .013
WHEN HOLES AT
HIGH LIMIT

The foregoing illustrates the interrelationship of size and location tolerances which is utilized in position dimensioning and tolerancing.

Although in this example we have used only two of the holes, the same reasoning applies to *all* the holes in the pattern; similarly, each individual hole could be offset within its tolerance zone in any direction around 360° and provide an acceptable situation.

It should be noted that a functional or fixed pin gage such as the one used here to explain the position theory can be used *only* to check the *positional* location of the holes. *Positional* tolerance can be added as the holes increase in size or depart from MMC size within their size tolerance range. Hole size tolerance, however, must be held within the tolerances specified on the drawing and must be checked individually and separately from the positional check.

MATING PARTS—FLOATING FASTENER

Position tolerancing techniques are most effective and appropriate in mating part situations. The illustrations on page 109, in addition to demonstrating the calculations required, also emphasize the importance of decisions at the design stage to recognize and initiate the position principles.

The mating parts shown in the illustration are to be interchangeable. Thus the calculation of their position tolerances should be based on the two parts and their interface with the fastener in terms of MMC sizes.

The two parts are to be assembled with four .190 screws. The holes in the two parts are to line up sufficiently to pass the four screws at assembly. Since the four screws ("fasteners") are separate components, they are considered to have some "float" with respect to one another. The colloquial term, "floating fastener" application, has been popularly used to describe this situation.

The calculations are shown in the upper right corner of the illustration. Also, note that, in this case, the same basic dimensions and position tolerances are used on both parts. They are, of course, separate parts and are on separate drawings.

The position tolerance calculations are based on the MMC sizes of the holes and the screws. The maximum material basis then sets the stage for maximum producibility, interchangeability, functional gaging (if desired), etc., at production. As seen from the illustration, part acceptance tolerances will increase as the hole sizes in the parts are actually produced and vary in size as a departure from MMC. From the .016 diameter tolerance calculated, the tolerance may increase to as much as .022 dependent upon the actually produced hole size. It should be noted that clearance between the mating features (in this case hole and screw) is the criterion for establishing the position tolerances.

Simultaneously with these production advantages, the design is protected since it has been based upon the realities of the hole and screw sizes as they interrelate at assembly and in their function. Thus, as parts are produced, assembly is ensured, and the design function is carried out specifically as planned.

A possible functional gage is also shown in the illustration on page 109. The .190 gage pin diameters are determined by the MMC size of the hole, .206, minus the stated position tolerance of .016. In our example, the same functional gage can be used on both parts. Functional gages are, of course, not required with position application, but they do, however, provide an effective method of evaluation where desired.

Referring to the position tolerance calculations, if more than two parts are assembled in a floating fastener application, we must determine the position tolerance to ensure that any two parts and the fastener will mate properly. Calculate each part to mate with the fastener using the illustrated formula and MMC sizes.

Only the primary datums (the interface surfaces) are shown on the illustrations for simplicity of explanation of this method. The hole pattern alignment is emphasized between the two parts. A complete specification of these requirements on both parts would include two additional datums to locate the holes and the pattern from the part outside surfaces (datums).

The clearance holes on these parts are all specified as same size. Where they are specified as *different sizes,* each part can be calculated separately for the allowable positional displacement based on the difference between the MMC of the hole and fastener. If one part in our example had .203 ± .003—4 HOLES specified, the method below would be used:

```
MMC hole          0.200
MMC fastener   (−) 0.190        | ⊕ | ⌀ .010 Ⓜ | A |
                  0.010

Other part
MMC hole          0.206
MMC fastener   (−) 0.190        | ⊕ | ⌀ .016 Ⓜ | A |
                  0.016
```

The position tolerance calculation method illustrated assumes the possiblity of a zero inter-ference-zero clearance condition of the mating part features at extreme tolerance limits. Additional compensation of the calculated tolerance values should be considered as necessary relative to the particular application.

Formulas used as a basis for the position floating fastener calculations are:

To calculate position tolerance with fastener and hole size known:

$$T = H - F$$

where T = tolerance, H = MMC hole, and F = MMC fastener.

Where the hole size or fastener size is to be derived from an established position tolerance, the formula is altered to:

$$H = T + F$$
$$F = H - T$$

MATING PARTS—FLOATING FASTENER

EXAMPLE

CALCULATIONS
MMC OF HOLE   = ⌀.206
MMC OF FASTENER
    SCREW MAX = (−) ⌀.190
POSITIONAL TOL   = ⌀.016
FOR ALL HOLES ON
BOTH PARTS

PART ACCEPTANCE TOL
PART HOLES AT LOW
LIMIT .206 (MMC)
POSITIONAL TOL = .016 DIA

PART HOLES AT HIGH
LIMIT .212 (LMC)
POSITIONAL TOL = .022 DIA

AT SIZE	TOL
.206 MMC	.016
.207	.017
.208	.018
.209	.019
.210	.020
.211	.021
.212 LMC	.022

GAGE FOR ABOVE PARTS

.190
4 PLACES { .206 MMC / (−).016 TOL / .190 PIN }

MATING PARTS—FIXED FASTENER

When one of two mating parts has "fixed" features, such as the threaded studs in this example, the "fixed fastener" method is used in calculating position tolerances.

The term "fixed fastener" is a colloquialism popularly used to describe this application. Both the term and the technique are applied to numerous other manufacturing situations such as locating dowels and holes, tapped holes, etc.

The advantages of the MMC principle as described in the foregoing "floating fastener" application also apply here. However, with a "fixed fastener" application, the difference between the MMC sizes of mating features must be divided between the two features, since the total position tolerance must be shared by the two mating features. In this example, the two mating features (actually four of each in each pattern) are the studs and the clearance holes. The studs must fit through the holes at assembly.

Again, we see that the clearance of the mating features as they relate to each other at assembly determines the position tolerances. When one feature is to be assembled *within* another on the basis of the MMC sizes and "worst" condition of assembly, the clearance, or total tolerance, must be divided for assignment to *each* of the mating part features. In this case, the derived .016 was divided equally, with .008 diameter position tolerance assigned to each mating part feature (stud and hole). The total tolerance of .016 can be distributed to the two parts as desired, so long as the total is .016 (e.g., .010 + .006, .012 + .004, etc.). This decision is made at the design stage, however, and must be fixed on the drawing before release to production.

Application of the MMC principle to situations of this type guarantees functional interchangeability, design integrity, maximum production tolerance, functional gaging (if desired), and uniform understanding of the requirements.

As the part features of both parts are produced, any departure in size from MMC will increase the calculated position by an amount equal to that departure. Thus, for example, the position tolerance of the upper part could possibly increase up to .014 and that of the lower part up to .013 dependent upon the amount of departure from their MMC sizes. However, parts must actually be produced and sizes established before the *amount* of increase in tolerance can be determined.

Functional gages (shown below each part in the illustration) can be used for checking, and, although their use is not a must, they provide a very effective method of evaluation if desired. Note that the functional gages resemble the mating parts; as a matter of fact, functional gages simulate mating parts at their worst condition.

The functional gage pins of the upper part are determined by the MMC hole size minus the stated position tolerance. Gage tolerances are not shown, although they may be imagined to be on the order of $.1981 \, ^{+.0002}_{-.0000}$ for pin size, and $\pm.0002$ on between pin locations. Local gage practices would prevail.

The functional gage on the lower part of the illustration contains holes instead of pins. The gage hole sizes are determined by the MMC (O.D.) size of the .190 pins plus the stated position tolerance. The tolerances are similar to those of the above pin gage. Tolerances on the order of $.1979 \, ^{+.0000}_{-.0002}$ for hole size, and $\pm.0002$ between holes could be applied, depending on local gage practices.

PART

1.500

1.250

4X ⌀.206 +.006 / −.000

⊕ | ⌀.008 Ⓜ | A

−A−

GAGE

1.500

1.250

.198
4 PLACES { .206 MMC / (−).008 TOL / .198 PIN }

PART

1.500

1.250

4X .190 −32 UNF−2A

⊕ | ⌀.008 Ⓜ | A

MAJOR ⌀

−A−

GAGE

1.500

1.250

.198
4 HOLES { .190 MMC / (+).008 TOL / .198 HOLE }

CALCULATIONS
MMC OF HOLE =⌀.206
MMC OF STUD =(−)⌀.190
2 ⌋.016
POSITIONAL TOL ⌀.008
FOR ALL HOLES
& STUDS

PART ACCEPTANCE TOL

PART HOLES AT LOW
LIMIT .206 (MMC)
POSITIONAL TOL = .008 DIA

PART HOLES AT HIGH
LIMIT .212 (LMC)
POSITIONAL TOL = .014 DIA

SIZE ⌀	⌀ TOL
.206 MMC	.008
.207	.009
.208	.010
.209	.011
.210	.012
.211	.013
.212 LMC	.014

SIZE ⌀	⌀ TOL
.190 MMC	.008
.189	.009
.188	.010
.187	.011
.186	.012
.185 LMC	.013

PART ACCEPTANCE TOL

PART STUDS AT HIGH
LIMIT .190 (MMC)
POSITIONAL TOL = .008 DIA

PART STUDS AT LOW
LIMIT .185 EST. (LMC)
POSITIONAL TOL = .013 DIA

It should be noted that the term MAJOR $\emptyset$ is used beneath the position callout on the lower part. In the absence of this special notation of exception, ANSI Y14.5, the Pitch Diameter Rule would have invoked the tolerance on the basis of the pitch diameter of the threads. The major diameter (or O.D.) of the thread was the desired criterion in this example. See POSITION—EXTENDED PRINCIPLES section for additional examples of fixed fastener applications.

The calculations on these parts illustrate a balanced tolerance application in which the total permissible position tolerance of the two parts is equally divided, for example, .008 on each part. The total position tolerance can, however, be distributed as desired, as discussed earlier.

If more than two parts are assembled in a fixed fastener application, each part containing clearance holes must be calculated to mate with the part with the fixed features.

The position tolerance calculation method illustrated assumes the possibility of a zero interference-zero clearance condition of the mating part features at extreme tolerance limits. Additional compensation of the calculated tolerance values should be considered as necessary relative to the particular application.

Formulas used as a basis for the position fixed fastener (or locator) calculations are:

$$T = \frac{H - F}{2}$$

MMC hole = H
MMC fastener = F
(or pin, dowel, etc.)
Tolerance = T

Where the hole size or fastener (or pin, dowel, etc.) size is to be derived from an established tolerance, the formula is altered to:

$$H = F + 2T$$
$$F = H - 2T$$

The illustration on page 113 shows position tolerancing applied to two mating parts with a circular hole pattern. The same reasoning applies here as in the preceding examples except that the basic dimensions are angular (45° angles, 8 places) and a diameter (the 1.500 diameter).

These two parts again are of the fixed fastening type, the studs of the lower part being the fixed elements. To determine the positional tolerances for each part, the MMC of the hole and the MMC of the stud are used to determine the total positional tolerance. This is divided by two to give the positional tolerance value for each part. The total value may be divided as desired, as previously described.

Note again how the positional tolerance *increases* as the holes in the upper part and the studs in the lower part depart from their MMC sizes, that is, when the holes get larger and the pins get smaller during the production process.

Functional gages are shown in the illustration for each of the parts. Note that the pins in the upper gage are calculated to the MMC or low limit of the holes in the part (which is .187 in this case) *minus* the positional tolerance (.0025), resulting in the .1845 gage pin size.

8X ⌀.190 ±.003

| ⊕ | ⌀.0025 Ⓜ | A |

8X 45°

−A−

⌀.187 MMC
(−).0025 TOL
⌀.1845 PIN
8X ⌀.1845

⌀1.500

CALCULATIONS
MMC OF HOLE = ⌀.187
MMC OF STUD = (−)⌀.182
2 |.005
POSITIONAL = ⌀.0025
TOL FOR ALL
HOLES AND STUDS

PART ACCEPTANCE TOL

PART HOLES AT LOW
LIMIT .187 (MMC)
POSITIONAL TOL = ⌀.0025

PART HOLES AT HIGH
LIMIT .193
POSITIONAL TOL = ⌀.0085

45°

GAGE

⌀1.500

⌀ SIZE	⌀ TOL
.187 MMC	.0025
.1875	.003
.188	.0035
.189	.0045
.190	.0055
.191	.0065
.192	.0075
.193 LMC	.0085

8X ⌀.180 ±.002

| ⊕ | ⌀.0025 Ⓜ | A |

8X 45°

−A−

⌀.182 MMC
(+).0025 TOL
⌀.1845 HOLE
8X ⌀.1845

⌀1.500

PART ACCEPTANCE TOL

PART STUDS AT HIGH
LIMIT .182 (MMC)
POSITIONAL TOL = ⌀.0025

PART STUDS AT LOW
LIMIT .178
POSITIONAL TOL = ⌀.0065

8X 45°

GAGE

⌀1.500

⌀ SIZE	⌀ TOL
.182 MMC	.0025
.1815	.003
.181	.0035
.180	.0045
.179	.0055
.178 LMC	.0065

The lower gage is calculated in reverse, using the MMC on high limit of the stud, .182, plus the positional tolerance (.0025), resulting in the .1845 gage hole size.

These calculations illustrate a balanced tolerance application in which the total permissible position tolerance of the two parts is equally divided, for example, .0025 on each part. The total position tolerance can, however, be distributed as desired, for example, .002 on one part, .003 on the other, etc., so long as it totals the tolerance calculated (in this case .005).

The position tolerance calculation method illustrated here and in preceding examples assumes the possibility of a zero interference-zero clearance condition of the mating part features at extreme tolerance limits. Additional compensation of the calculated tolerance value should be considered as necessary relative to the particular application.

A primary datum (A) only is shown in the examples for simplicity of explanation. A secondary datum (i.e., the outside diameter) would be required to complete the specifications on both parts.

RELATION TO IMPLIED DATUM SURFACES*

Datum planes or surfaces as the basis for position relationships may be either *implied,** with no datum callouts on the drawing, or *specified* on the drawing. It should be made clear, however, that implied datums of any variety tend toward ambiguity. Caution should be exercised in their use. This discussion is intended to be an interpretation of principles applied on less critical requirements or where dimensional construction implies surface datums reasonably clearly. Datum precedence is, of course, left in doubt.

This illustration shows a part with three holes arranged in a position pattern. Datums are not specified on the drawing, because the datums are *implied* by the surfaces *from which* the position pattern dimensions are taken. In this case, the .750 dimensions obviously determine the surfaces which are to be used as datums and tie the surfaces to the position hole pattern.

The extremities or high points of the datum surfaces from which the .750 dimensions are taken establish the datum planes from which the position hole pattern is oriented. These datum planes are functional to the part requirement and are also used for tooling or fixturing reference and in establishing measuring planes to inspect the part.

The position tolerance zone is .010 when the holes are produced at MMC or the low limit of .245. The tolerance zone increases up to .020, if the holes are produced larger to the high limit of .255. Note that the position tolerance applies while the part is in contact with the datum surface; thus the position tolerance stated also controls the pattern location from the edges and provides the tolerance for the .750 pattern locating dimensions.

Where ambiguity exists in the use of implied surface datums or where datum precedence is required, datums should be specified. Implied datums, other than on outside surfaces of noncylindrical shapes, should normally not be permitted or used.

---

*Implied datums are no longer approved per the latest standard ANSI Y14.5–1982; all datums must now be specified. This example and coverage have been retained for informational purposes and to assist in interpreting drawings prepared in accordance with earlier standards.

AS DRAWN

.250 ±.005–3 HOLES
⊕ | Ø.010 Ⓜ

1.500

3.00 ±.01

.750

1.750

.750

3.25
±.01

←.200 ±.005

MEANING

SIZE	Ø TOL
.245 MMC	.010
.246	.011
.247	.012
.250	.015
.255 LMC	.020

.245/.255
3 HOLES

HOLE (3) LOCATED AT TRUE POSITION WITHIN .010 DIA. IF HOLE AT .245 MMC, .020 DIA IF HOLE AT .255 HIGH LIMIT.

90°

90°

1.500

90°

90°

90°

.750

.750

1.750

.195
.205

90°

IMPLIED DATUM SURFACES
& PLANES
(DATUM PLANES DETERMINED
BY EXTREMITIES OF DATUM
SURFACES)

See note at the bottom of the previous page.

RELATION TO SPECIFIED DATUM SURFACES

The illustration on the right shows the same part as that shown on the preceding page, but the datums are identified with the A, B, and C datum feature symbols, and form and orientation tolerances are specified.

Where part function, and thus the stated drawing requirements, are more critical, specified datums and greater geometric control are essential to the design.

In this example, it was considered also necessary to control the accuracy of the datum surfaces in their specific relationship to each other. To accomplish this, identification of the specific surfaces as datum references was required. Further, since the hole position pattern was critical in its orientation to the surfaces, datum identification was required. With specification of the datums, precedence of the datum surfaces is established, and the part and the hole pattern are stablilized relative to the datum reference frame. Datum precedence is also established.

Datum surface A (top surface of the part) is to be held to a flatness of .001 total. Datum surface B is to be perpendicular to datum plane A within .001 total. Datum surface C is to be perpendicular to datum plane A within .001 total and also perpendicular to datum plane B within .002 total. Note that in this example we have also used combined datum feature symbols and feature control frames.

The extremities or high points of the datum surfaces from which the .750 dimensions are taken establish the secondary and tertiary datum planes (B and C). The primary datum plane (A) is established by the extremities of datum surface A. The part orientation with respect to the position pattern is thus fixed. These datum planes are functional to the part requirement and are also used for tooling or fixturing reference.

The position tolerance zone is .010 when the holes are produced at MMC or the low limit of .245. The tolerance zone increases up to .020 if the holes are produced larger to the high limit of .255. Note that the position tolerance applies while the part is in contact with the datum surfaces according to their stated precedence or sequence. The position tolerance stated also controls the pattern location from the edges and provides the tolerance for the .750 pattern locating dimensions.

In this example all geometric controls are specifically stated, removing all doubt as to design intent and follow-through manufacture and inspection requirements. The tools provided by specified datums and greater geometric control can be very effectively applied and will protect design integrity.

It should be noted that the form and orientation controls shown are not necessarily required when specifying the datum planes (A, B, C). The designer should consider the precision of the surfaces being used as datum features (as controlled by size), and then only as necessary, per the design requirements, add any form or orientation controls.

POSITION TOLERANCE—SPECIFIED DATUMS

AS DRAWN

SIZE	Ø TOL
.245 MMC	.010
.246	.011
.247	.012
.250	.015
.255 LMC	.020

HOLE (3) LOCATED AT POSITION WITHIN .010 DIA IF HOLE AT .245 MMC, .020 DIA IF HOLE AT .255 LMC.

MEANING

DATUM PLANE B

.245/.255 3 HOLES

DATUM PLANE C

DATUM PLANE A

DATUM SURFACE "B" PERPENDICULAR TO "A" WITHIN .001

DATUM SURFACE "C" PERPENDICULAR TO "A" WITHIN .001 & PERPENDICULAR TO "B" WITHIN .002

DATUM SURFACE "A" FLAT WITHIN .001 TOTAL

(DATUM PLANES DETERMINED BY EXTREMITIES OF DATUM SURFACES)

COMBINATION COORDINATE AND POSITION TOLERANCING*

When the location of a pattern of features from datum surfaces is less important than the accuracy required within the pattern of features, both directly coordinated toleranced dimensions and position dimensions may be used. This approach may be used for mounting individual electrical or mechanical components, with the relationship of the mounting holes of the individual components as the most important consideration. When such combination dimensioning is used, the directly toleranced dimensions are to be interpreted as referring to theoretical centerlines (or center planes) in each pattern. Each individual feature is then permitted to deviate from its position in the pattern, within the position tolerance for that feature. In this case, a primary datum (A) is specified, with the less important outer edges implied as the secondary and tertiary datums.

In the upper right pattern of mounting holes in the example, the theoretical center of the pattern is located by toleranced dimensions from the datum surfaces. This center is the axis for the 1.990 basic diameter. The individual theoretical centers of the eight .188 holes are established by intersection of the 45° basic angles at the 1.990 basic diameter.

The actual centers of the individual holes are specified to be at position within a .005 diameter tolerance zone when the hole is at .188 MMC. The tolerance zone increases to .009 (the position tolerance of .005 plus the hole tolerance of .004) as the hole is produced at the high limit or largest size of .192 (LMC).

The lower rectangular pattern of holes shows the center planes located by toleranced dimensions from the datum surfaces and basic dimensions within the pattern. The meaning is shown in the lower portion of the page.

The position tolerance zone is .005 diameter total when the holes are at MMC of .250 and .009 diameter total when the holes are at .254, or the high limit (LMC).

Note that in each hole pattern the pattern as a unit may vary within the more lenient pattern locating coordinate tolerance zone, while the hole-to-hole location within each pattern must be maintained within the position tolerance zone.

As is seen in the details of the lower illustration at right, the position tolerance zone (center or axis) must lie within, or on, the pattern location zone with the produced hole axis permitted to fall anywhere within the position tolerance zone.

Likewise seen is the restriction implied for the perpendicularity or attitude of each hole within (only) the position tolerance zone; the pattern location or rotation is, however, permitted the wider coordinate tolerance. Departure from MMC hole size then increases the tolerances as previously described.

---

*This method is no longer approved per the latest standard ANSI Y14.5M–1982; it is now considered ''former practices.'' This example and coverage have been retained for informational purposes and to assist in interpreting drawings prepared in accordance with earlier standards.

AS DRAWN

.188 $^{+.004}_{-.000}$ 8 HOLES

⊕ | ⌀.005 Ⓜ | A

1.990

2.008 ±.015

-A-

45°
8 PLACES

1.000

1.500

3.978
±.015

.410
±.015

.525 ±.015

.250 $^{+.004}_{-.000}$ 4 HOLES

⊕ | ⌀.005 Ⓜ | A

MEANING

POSITION TOL ZONES
.005⌀ (AT HOLE MMC)
.009⌀ (AT HOLE LMC)

HOLE

.030 SQUARE
TOL ZONE

3.993
3.963

1.993
2.023

IMPLIED DATUM PLANES

PERMISSIBLE HOLE
AXIS VARIATION

.005⌀ TOL ZONE
AT MMC

DATUM PLANE A

90°

TYPICAL HOLE
CROSS SECTION

.030 PATTERN
LOCATING
TOL ZONE

1.000

1.500

90°

1.925
1.895

.425
.395

POSITION TOL ZONES
.005⌀ (AT HOLE MMC)
.009⌀ (AT HOLE LMC)

POSITIONAL TOL ZONE (CENTER OR
AXIS) MUST LIE WITHIN OR ON
PERIPHERY OF RESPECTIVE PATTERN
LOCATING ZONE. HOLE CENTER
(AXIS) MAY VARY WITHIN POSITION
TOL ZONE.

.030 SQUARE
TOL ZONES

HOLE CENTER (AXIS)

IMPLIED
DATUM
PLANES

.510

.540

1.510

1.540

See note at the bottom of the previous page.

RELATION TO DATUM SURFACES
COMPOSITE POSITION TOLERANCING

When the location of a pattern of features from datum surfaces is less important than the accuracy required *within* the pattern of features, composite position tolerancing may be used. Composite position tolerancing also extends the use of specified datum relationships and geometric tolerance control.

Composite position tolerancing incorporates a dual feature control frame with two positional controls. One, the upper entry in the symbol, specifies the applicable datums and the pattern locating position tolerance. The lower entry specifies the applicable datum and the feature relating position tolerance. A single position tolerance symbol is used.

The composite position tolerance method utilizes the full advantages of MMC and extends the principles to control of patterns of features as well as of the individual feature interrelationship.

In the upper right pattern of eight mounting holes in the example, the theoretical center of the pattern is located by the 2.008 and 3.978 basic dimensions and is related to datums A, B, and C. This center is the axis for the 1.990 basic diameter. The individual theoretical centers of the eight .188 holes are established by the intersection of the 45° basic angles at the 1.990 basic diameter. The pattern as a unit, yet actually determined by the holes themselves, is to be located within a position tolerance of .030 when the holes are at MMC. As the individual holes in the pattern depart from MMC toward LMC, additional tolerance to that hole (and thus to the pattern) is acquired equal to that departure.

Within the hole pattern itself, the feature relating position tolerance is established at .005 diameter. Then, as previously described, each hole in the pattern may increase its position tolerance an amount equal to the departure from MMC as it is produced to a maximum of .009 at LMC. Note that the hole-to-hole interrelationship in the pattern, as well as the relationship to datum A, is maintained. The attitude (perpendicularity) of each individual hole must be within the .005 diameter tolerance zone as well as within position of .005 at MMC.

As can be seen in the typical hole cross-sectional view, the axes of both the large (pattern) and small (feature) tolerance zones are parallel. The axes of the holes must lie within *both* the larger and smaller tolerance zones. Portions of the smaller zones may fall outside the peripheries of the larger tolerance. However, this portion of the smaller zone is not usable since the axis of each hole must fall within both zones.

The lower left four hole pattern follows the same reasoning as described above and as seen in the illustration.

AS DRAWN

8X Ø.188 +.004 / −.000

⊕	Ø.030 Ⓜ	A	B	C
	Ø.005 Ⓜ	A		

−B−

2.008

Ø1.990

−A−

8X 45°

3.978

1.000

1.500

.410

−C−

.525

4X Ø.250 +.004 / −.000

⊕	Ø.030 Ⓜ	A	B	C
	Ø.005 Ⓜ	A		

MEANING

FEATURE RELATING
POSITION TOL ZONES
.005Ø (AT HOLE MMC)
.009Ø (AT HOLE LMC)

HOLE
CENTER
(AXIS)

PATTERN LOCATING
POSITION TOL ZONE
.030Ø (AT HOLE MMC)
.034Ø (AT HOLE LMC)

3.978

2.008

DATUM B

DATUM C

PERMISSIBLE HOLE
AXIS VARIATION

.005Ø TOL ZONE
AT MMC

DATUM PLANE
A

TYPICAL HOLE
CROSS SECTION 90°

.030Ø PATTERN
LOCATING TOL ZONE
AT MMC

ACTUAL HOLE LOCATION
MAY VARY FROM
PERPENDICULAR TO
PRIMARY DATUM PLANE A
WITHIN POSN. TOL ONLY. HOLE
PATTERN AS A UNIT (AS
DETERMINED BY THE HOLES)
MAY VARY WITHIN PATTERN
LOCATING TOL ZONE.

HOLE CENTER (AXIS)

FEATURE RELATING-POSITION
TOL ZONES
.005Ø (AT HOLE MMC)
.009Ø (AT HOLE LMC)

PATTERN LOCATING-POSITION
TOL ZONES
.030Ø (AT HOLE MMC)
.034Ø (AT HOLE LMC)

THE AXIS OF THE HOLES MUST
LIE WITHIN BOTH THE PATTERN
LOCATING TOL ZONES (.030Ø AT
MMC) AND THE FEATURE
RELATING TOL ZONES (.005Ø
AT MMC)

1.000

1.500

1.500

90° (4)

.410

.525

1.000

DATUM
PLANE B

DATUM PLANE C

See following pages for gaging.

RELATION TO DATUM SURFACES
COMPOSITE POSITION TOLERANCING—FUNCTIONAL GAGES

AS DRAWN   **GAGES**

COMPOSITE POSITION TOLERANCING—FUNCTIONAL AND PAPER GAGING, COMPUTATION

The previous page illustrates functional gage principles for the four hole pattern of the part on page 121. Other methods—paper gaging and mathematical computation—are represented by a sample situation on page 125.

Paper gaging methods are shown to demonstrate actual usable techniques which can also be used to quantify position tolerance principles pictorially. The mathematical method may be used in lieu of paper gaging or functional gaging directly as backup verification of border-line cases, or for further analytical work.

PAPER GAGING

Paper gaging is accomplished through plotting an enlarged scale of coordinately measured feature positions onto a piece of standard graph paper and then plotting the resulting differentials (actual position versus true position) to a selected scale (e.g., one square = .001) with a dot on the graph. An overlay chart (gage) of tracing paper or other transparent material containing a series of graph-scale circles of desired increments is placed over the graph to depict the position tolerance zones. Note that the paper gaging method simulates part function and functional gaging. However, the individual tolerance zones are each assumed to be represented by the one exact (true) position on the graph. The exact (basic) dimensions of the pattern are assumed as 0 in the X and Y directions.

Step 1. The four .250 holes are located both as a pattern (ϕ .030) and hole-to-hole within the pattern (ϕ .005). In open set-up inspection methods, two steps are required to determine whether both requirements have been met. Step 1, using coordinate measuring and paper gaging, will determine whether the hole pattern as a unit (as based upon individual hole location) has met the ϕ .030 tolerance requirement. The part is set up according to the datum surfaces and is measured to holes 1, 2, and 3 (see illustration on page 125); only three of the four holes are necessary for this evaluation.

The resulting X and Y measurements are compared to the specified coordinate dimensions on the drawing resulting in a differential (off position) value. This differential is plotted on the paper gage graph according to the above stated scale (i.e., one square = .001).

When the actual location of the three holes are plotted, an overlay gage, with the circles, the scale of the graph plot, and a representation of the tolerances, is placed over the graph plot. The center of the overlay must be placed on the center of the plot. If the plotted centers fall within the position zone (in this case the ϕ .030 zone), the pattern, via the holes, meets the requirement. Since hole #1 in our hypothetical example exceeds the ϕ .030 tolerance zone, MMC principles may be invoked. The size of hole #1 is found to be .254 which is .004 departure from MMC; thus that hole has ϕ .034 tolerance, and as seen in the illustration on the following page, is now acceptable. The hole pattern is also thus found acceptable within the ϕ .030 tolerance.

Step 2. To evaluate the accuracy of the four holes in the pattern (part shown on page 121) relative to the individual .005 position tolerances, it is necessary to consider the hole-to-hole relationship in the pattern, exclusively. The four-hole grid pattern from which the individual hole positions can be compared is established by selecting two of the holes as a basis. In our example, hole #1 is selected as the origin for the X- and Y- coordinate measurements.

Since one additional hole must be selected to give orientation or square-up to the pattern, hole #3 is selected for the X orientation.

From the part orientation now established, each hole is measured in X and Y from this set-up. From the illustration on page 125, resultant differentials are derived from our hypothetical measurements. Note, of course, that hole #1 is zero in X and Y (as the origin for measurement), and hole #3 is zero in X as the square-up for the basic pattern.

The differentials are plotted on the graph paper to the desired scale (in this case every five squares = .001) using an origin on the graph as the hole #1 position. Each hole is plotted in the appropriate value and quadrant in X and Y from the hole #1 position.

With the holes plotted on the graph, the overlay gage with the circles representing the tolerances scaled to the graph plot is placed over the graph. The overlay is moved at random to try to encompass all the plotted hole centers simultaneously within the stated tolerance .005 circles. After trial and error, let us assume that the illustrated location of the overlay gage can successfully encompass holes #1, #2, and #3 but hole #4 is outside the .005 circle. MMC principles may now be invoked. By checking the size of hole #4, it is determined to be .2515, which is a departure of .0015 from MMC; thus hole #4 has .0065 position tolerance, and as is seen in the illustration, is now acceptable.

All holes in the pattern have met their position tolerance requirements. It is extremely important to note that hole #1, even though used as a zero location origin, is actually assumed as an equal partner in the pattern and must likewise be treated as imperfect in its relationship to its desired position.

Paper gaging simulates hard gaging and part function and thus is an effective technique. This method is further advantageous in that it visually detects error trends through periodic inspections, gives a permanent record, and requires no gaging tolerance.

COMPUTATION

Mathematical verification is possible using a calculator or computer programs, proving the holes within the ⌀ .030 and ⌀ .005 tolerance zones.

Step 1 Mathematically proven using a calculator or computer program (similar to pages 281–284).

- Hole #1 calculates to ⌀ .0328024 − .004 (bonus tol) = ⌀ .0288024. Hole #1 is *good* using the .004 bonus tolerance (less than ⌀ .030).

- Hole #2 calculates to = ⌀ .0268328. Hole #2 is *good* (less than ⌀ .030).

- Hole #3 calculates to = ⌀ .0233238. Hole #3 is *good* (less than ⌀ .030).

Step 2 Mathematically proven using a suitable computer program (similar to pages 284–289).

- Hole—hole location *is* within tolerance as calculated.

- Hole—hole position tolerance calculated for *all four holes,* considering bonus tolerance of hole #4 (⌀ .0015), is ⌀ .0047472. That is, the smallest circle (tolerance zone) which will encompass all the four hole locations (including bonus tol ⌀ .0015 of hole #4) is ⌀ .0047472. This is less than ⌀ .005, the stated tolerance. Therefore, all four holes are acceptable.

STEP 1

OVERLAY GAGE
(TOL ZONE)

.034
.032
.030
#1

GAGE CENTER
(TRUE POSITION)
#2

DATUM B

#3

GRAPH

SCALE ■ = .001

0

DATUM C

1.000
1.500
.410
-C-
.525

3   4
1   2

4X ⌀.250 +.004 -.000

⊕ | ⌀.030 Ⓜ | A | B | C
  | ⌀.005 Ⓜ | A

HOLE

		X	Y	HOLE SIZE	TOL
①		.538 (Actual)	.420	⌀.254 (Actual)	⌀.030
		.525 (Drawing)	.410	.250 (MMC)	+.004
		.013 (+X)	.010 (+Y)	⌀.004	⌀.034
②		1.525	.422		⌀.030
		1.519	.410		
		.006 (−X)	.012 (+Y)		
③		.525	1.910		⌀.030
		.515	1.904		
		.010 (−X)	.006 (−Y)		

STEP 2

(OVERLAY GAGE)

.005
.0055
.006
.0065

OVERLAY GAGE
(TOL ZONE)

#1

.005
.0055
.006
.0065

0

SCALE ▨ = .001
■ = .0002

GRAPH

#2
#3
#4

1.000
1.500
.410
-C-
.525

3   4
1   2

4X ⌀.250 +.004 -.000

⊕ | ⌀.030 Ⓜ | A | B | C
  | ⌀.005 Ⓜ | A

		X	Y	HOLE SIZE	TOL
①		X = O (ORIGIN)	Y = O (ORIGIN)	GO/NO GO	⌀.005
②		1.0000	(−).0042	GO/NO GO	⌀.005
		.9988	.0000		
		.0012 (−X)	.0042 (−Y)		
③		X = O (SQUARE-UP)	1.5000	GO/NO GO	⌀.005
			1.4960		
			.0040 (−Y)		
④		1.0032	1.5000	⌀.2515 (Actual)	⌀.005
		1.0000	1.4965	⌀.2500 (MMC)	+.0015
		.0032 (+X)	.0035 (−Y)	⌀.0015	⌀.0065

COMPOSITE POSITION TOLERANCING

The composite position tolerancing principle can be applied in numerous ways to meet similar, yet different, design requirements from that shown on previous pages. For example, where the two patterns of features are in fact *one* pattern, or are to be treated as *one* pattern for both the pattern locating and feature relating (upper *and* lower entries), the method shown in the upper example could be used. One callout addressed to both patterns and the number designation 12X (twelve holes) will convey the one pattern meaning.

Functional gaging, coordinate measuring, and evaluation of the upper entry, pattern locating tolerance, would be exactly as shown in previous pages. The lower entry, feature relating tolerance, would also be handled as in previous pages except that it would be put into a *single gage* (accommodating all twelve holes at once) or into a *single measuring process* where all twelve holes are measured and evaluated as *one* pattern.

The lower example illustrates the possibility where one pattern (the four ϕ .250 holes) is to serve as the datum reference. Although feasible, this method includes some complications in verification. Where the design requirement is that the diametrical pattern (the eight ϕ .188 holes) be located with respect to the four holes (designated datum D) location, this method can be used. Production is so guided. However, verification from the four holes requires either some form of functional gage which can pick up the four holes simultaneously (and thus their functional centroid), or open set-up measuring by coordinate methods selecting two of the four holes or mathematically deriving the centroid of the four holes via computer. Methods shown previously can then be applied with some trial and error attempts possibly necessary.

If a functional gage method is to be used in this application, due consideration must be given to the Datum Virtual Rule as it would apply to the datum D pattern gage pins (i.e., ϕ .245). Should the design require more stringent control or relationship of the eight hole pattern to the datum D pattern, zero (.000) tolerancing of the feature relating hole pattern could be considered (see below). In such a case, the Datum Virtual Condition Rule would not apply since the gage pins would be at MMC size.

i.e.

4X

$\oplus$	ϕ .030 Ⓜ	A	B	C
	ϕ .000 Ⓜ	A		

Consideration could be given to whether selection of two of the four holes might better serve as secondary and tertiary datum features (arbitrarily selected by design). This method would remove most of the complexities of using a group of features as a single datum. If RFS reference to the datum as a group is considered, the foregoing method (select two holes) should seriously be considered. Otherwise, the complexities become deeper and can only be resolved by sophisticated analytical or computer methods.

See later sections in this text for further discussion on two hole datum applications, coordinate measuring, relationship to gaging, and mathematical (computer) simulation of hole patterns.

COMPOSITE POSITION TOLERANCING—ONE PATTERN

COMPOSITE POSITION TOLERANCING—HOLE PATTERN AS DATUM

MMC RELATED TO MMC DATUM FEATURE

When a pattern of holes is dimensioned relative to the location of another hole, this hole is identified as a datum and the hole pattern is located dimensionally with respect to it.

In the example, the four holes are related to the center datum hole. As the position of the center datum hole shifts, the position of the 4 hole pattern itself must follow as dictated by the function of the part. Imagine that this part has a mating part with a shaft and four pins which must assemble with the five holes in the illustrated part.

Note that the .500 datum center hole is also located from datums (surfaces) A, B, and C. It is given a position tolerance of ϕ .010 at MMC and a refined orientation, or perpendicularity, tolerance of ϕ.003 at MMC. This properly controls the center hole relationship to the edge surfaces with a rather lenient position tolerance while the orientation to the primary datum A is maintained to closer tolerance. The center hole is identified as datum D so that the four .380 holes can be located with respect to it. Under MMC principles, the position and perpendicularity tolerances increase an amount equal to the hole size departure from MMC as shown in the illustration.

Wherever the datum hole D position varies in the design considerations, or in actual production, the four .380 hole pattern must follow. Note that the positional pattern dimensions originate at datum D to carry out this intent.

The four .380 holes are located by a position tolerance of ϕ .005 at MMC with respect to datum A (for orientation), datum D at MMC (virtual condition) (for location), and datum B (for rotation). Reference to the lower portion of the illustration under Meaning will assist in the understanding of the effect of the three datums and the importance of datum precedence. Although not illustrated, imagine the four .380 hole position tolerances have been calculated relative to a mating part using the "fixed fastener" method previously explained.

The four holes individually with respect to their own true positions can vary in location up to ϕ .005 at MMC. While at any size in a departure from MMC a hole can vary an additional amount (enlarges positional tolerance) equal to that departure. For example (as shown), if the hole size is produced at .381, the position tolerance becomes ϕ .006, at .382 it becomes ϕ .007, etc., up to ϕ .008 if the hole is produced at .383 LMC.

Since the datum feature is a size feature, its variation in size (.500 to .502) has an effect on the four hole pattern relationship which locates from it. That is, as the size of the datum feature increases, its relationship to the mating part corresponding feature (e.g., a shaft) changes; if the imagined mating part shaft and corresponding pins (at MMC) will insert into holes larger than those we used to calculate the position tolerances, greater latitude (off position) of the pattern as a unit can be realized. This latitude is, however, to the hole pattern as a unit and not relative to the four holes individually or hole-to-hole in the four hole pattern.

Please see the following gaging illustrations and text describing evaluation of this part with representative techniques.

AS DRAWN

2.75 ±.01

2.000

1.000

4X ⌀.380 $+.003\ -.000$

| ⊕ | ⌀.005 Ⓜ | A | D Ⓜ | B |

-B-

1.75 ±.01

.750

.375

.750

⌀.500 $+.002\ -.000$

.1400

.560 ±.005

| ⊕ | ⌀.010 Ⓜ | A | B | C |
| ⊥ | ⌀.003 Ⓜ | A | | |

-D-

-C-

-A-

SYMBOL MEANING

| ⊕ | ⌀.005 Ⓜ | A | D Ⓜ | B |

THIS FEATURE MUST BE AT TRUE POSITION

WITH RESPECT TO DATUMS A, D AT MMC (VIRTUAL COND.) AND B.

WITHIN .005 ⌀ TOL ZONE AT MMC

MEANING

HOLE AXIS

1.400

PRIMARY DATUM PLANE A

90°

SEC DATUM PLANE B

90°

.750

90°

TERTIARY DATUM PLANE C

.003⌀ TOL ZONE (AT MMC)

.010⌀ TOL ZONE (AT MMC)

DATUM HOLE POSITION TOL ZONE ⌀
.010 (AT MMC)
.012 (AT LMC)

⌀ SIZE	⌀ TOL
.500 (MMC)	.010
.5005	.0105
.501	.011
.5015	.0115
.502 (LMC)	.012

DATUM PERP. TOL ZONE ⌀
.003 (AT MMC)
.005 (AT LMC)

⌀ SIZE	⌀ TOL
.500 (MMC)	.003
.5005	.0035
.501	.004
.5015	.0045
.502 (LMC)	.005

TERTIARY DATUM B

2.000

1.000

SEC DATUM D ⌀.500 HOLE AT MMC

90°

PARALLEL

90°

PRIMARY DATUM PLANE A

.375

.750

90°

POSITION TOL ZONE ⌀ (4)
.005 (AT MMC)
.008 (AT LMC)

⌀ SIZE	⊕ TOL
.380 (MMC)	.005
.3805	.0055
.381	.006
.3815	.0065
.382	.007
.3825	.0075
.383 (LMC)	.008

.005⌀ TOL ZONE (AT MMC) (4)

.002⌀ TOL ZONE (AT LMC) (PATTERN, AS A UNIT, SHIFT) PERMISSIBLE

MMC RELATED TO MMC DATUM FEATURE
FUNCTIONAL GAGING PRINCIPLES

Functional gaging principles which can be applied to the preceding example (duplicated at the upper left of the opposite page) are shown in the illustration.

It must first be clarified that functional gaging is not required when position tolerancing is used. Other techniques (i.e., open set-up coordinate methods, optical methods, etc.) can also be used. For example, the paper gaging and computation methods illustrated earlier in this section could effectively be applied using the datum D feature as zero and orientation and rotation from datums A and B.

The purpose of this illustration is to depict representative functional gages and show actual methods as well as to demonstrate principle.

The illustration is intended to be self-descriptive and to explain the necessary details. Note that three gages (or operations) are required to fulfill the requirements.

Note specifically the clarity afforded by the clearly specified design requirements via the datums. Also, note the manner in which the gages pattern after the part function and simulate mating part relationships.

Also worthy of note is the manner in which the datum D pick-up pin size virtual condition is established based upon the Datum Virtual Condition Rule. The virtual condition applicable (.497) is derived from the smallest orientation or position tolerance controlling the hole (i.e., ϕ .003) as subtracted from MMC size .500.

AS DRAWN

2.75 ±.01

2.000

1.000

4X ∅.380 +.003 −.000

| ⊕ | ∅.005 Ⓜ | A | D Ⓜ | B |

−B−

1.75 ±.01

.750

.375

.750

∅.500 +.002 −.000

1.400

.560 ±.005

−C−

−A−

| ⊕ | ∅.010 Ⓜ | A | B | C |
| ⊥ | ∅.003 Ⓜ | A | | |

−D−

PERPENDICULARITY GAGE PIN CALCULATIONS (⊥)

DATUM A SIM

∅.497 GAGE PIN

90°

GAGE PIN CALCULATION
∅.500 +.002 −.000 HOLE
∅.500 MMC
(−).003∅ ⊥ TOL AT MMC
∅.497 GAGE PIN SIZE

FUNCTIONAL GAGES

.750

90°

SIM DATUM C

1.400

∅.490 GAGE PIN

90°

SIM DATUM B

90°

SIM DATUM A

∅.500 HOLE LOCATING (⊕) GAGE PIN CALCULATIONS
⌈ ∅.500 +.002 −.000 HOLE
∅.500 MMC
(−).010∅ ⊕ TOL AT MMC
∅.490 GAGE PIN SIZE ⌋

SIM DATUM B (PARALLEL)

.375

.750

90°

90°

∅.375 GAGE PIN (4)

1.000

2.000

DATUM D PIN ∅.497 (VIRTUAL COND.)

SIM DATUM A

PATTERN LOCATING (⊕) GAGE PIN CALCULATIONS
⌈ ∅.500 +.002 −.000 HOLE (DATUM D)
∅.500 MMC
(−).003∅ ⊥ TOL AT MMC
∅.497 DATUM PIN SIZE
(VIRTUAL CONDITION) ⌋

⌈ ∅.380 +.003 −.000 HOLE
∅.380 MMC
(−).005∅ ⊕ TOL AT MMC
∅.375 GAGE PIN SIZE ⌋

MMC RELATED TO RFS DATUM FEATURE

When a pattern of holes is dimensioned relative to the location of another hole, this hole is identified as a datum and the hole pattern is located dimensionally with respect to it.

In the illustrated example, the four holes are related to the center datum hole. As the position of the center datum hole shifts, the position of the four hole pattern itself must follow as dictated by the function of the part. Imagine that this part has a mating part with a shaft and four pins which must assemble with the five holes in the illustrated part.

Differing from the preceding illustration, assume there is a closer precision fit required between the shaft of the imagined part as it fits into the .500 hole. The four pins of the imagined part, however, are to relate to the four .380 holes on an MMC basis as in the preceding illustration. Therefore, since in this case the relationship between the four holes and their datum is critical or more precise, the datum D is referenced regardless of feature size (RFS).

Note that the .500 datum center hole is located from surface datums A, B, and C. It is given a position tolerance relative to these edges of ϕ .010 at MMC since the relationship to the edges on this basis can have a rather lenient position tolerance. The orientation of the datum hole relative to the primary datum A is, however, to be maintained to a closer tolerance. Since datum D position and orientation is controlled on an MMC basis, the tolerances (position and perpendicularity) increase an amount equal to the produced size departure from MMC shown in the illustration. It should be noted here, however, that the reference to datum D in the relationship of the four .380 holes is on an RFS basis. This means the four hole pattern takes reference from the exact center of the datum hole at whatever size it is produced (RFS) within .500 to .502.

Wherever the datum hole D position varies in the design considerations, or in actual production, the four .380 hole pattern must follow. Note that the positional pattern dimensions originate at datum D to carry out this intent.

The four .380 holes are located by a position tolerance of ϕ .005 at MMC with respect to datum A (for orientation), datum D at RFS (for location), and datum B (for rotation). Reference to the lower portion of the illustration under Meaning will assist in understanding the effect of the three datums and the importance of datum precedence. Although not illustrated, imagine that the four .380 hole position tolerances have been calculated relative to a mating part using the "fixed fastener" method previously explained.

The four holes individually, with respect to their own true or exact positions, can vary in location up to ϕ .005 at MMC. While at any size in a departure from MMC, a hole can vary an additional amount (enlarge positional tolerance) equal to that departure. For example (as shown), if the hole size is produced at .381, the position tolerance becomes ϕ .006, at .382 it becomes ϕ .007, etc., up to ϕ .008, if the hole is produced at .383 (LMC).

Although the datum feature is a size feature as in the preceding illustration, its variation in size (.500 to .502) will have *no* effect on the four hole pattern relationship. This is because the pattern relationship is to the center (axis) of the datum D, .500 hole, no matter to which size it is produced in its size tolerance range (i.e., RFS).

In this instance, it is seen that a more critical or precise relationship is maintained between the four holes and their datum. The design requirements and mating part functional interface determined this approach in our example.

Please see the following gaging illustrations and text describing evaluation of this part with representative techniques.

AS DRAWN

MEANING

MMC RELATED TO RFS DATUM FEATURE
FUNCTIONAL GAGING PRINCIPLES

Functional gaging principles which can be applied to the preceding example (duplicated at the upper left of the opposite page) are shown in the illustration.

It must first be clarified that functional gaging is not required when position tolerancing is used. Other techniques (i.e., open set-up coordinate methods, optical methods, etc.) can also be used. For example, the paper gaging and computation methods illustrated earlier in this section could effectively be applied using the datum D feature as zero and orientation and rotation from datums A and B.

The purpose in this illustration is to depict representative functional gages to show actual methods as well as to illustrate principle. Note that functional principles, identical to the preceding part, can be used except for the RFS pickup of the datum feature.

The illustration is intended to be self-descriptive and to explain the necessary details. Note that three gages (or operations) are required to fulfill the requirements.

Note specifically the clarity afforded by the clearly specified design requirements via the datums. Also, note the manner in which the gages pattern after the part function and simulate mating part relationships.

In this example, the use of the RFS datum has tightened requirements. Costs in manufacturing will probably be higher than with total use of MMC as in the preceding example; yet, functional principles can be applied as shown.

AS DRAWN

FUNCTIONAL GAGES

PERPENDICULARITY GAGE
PIN CALCULATIONS (⊥)

GAGE PIN CALCULATION

$\emptyset.500 \begin{smallmatrix} +.002 \\ -.000 \end{smallmatrix}$ HOLE
$\emptyset.500$ MMC
(−).003∅ ⊥ TOL AT MMC
$\emptyset.497$ GAGE PIN SIZE

$\emptyset.500$ HOLE LOCATING (⊕)
GAGE PIN CALCULATIONS

$\emptyset.500 \begin{smallmatrix} +.002 \\ -.000 \end{smallmatrix}$ HOLE
$\emptyset.500$ MMC
(−).010∅ ⊕ TOL AT MMC
$\emptyset.490$ GAGE PIN SIZE

PATTERN LOCATING
GAGE PIN CALCULATIONS

$\emptyset.380 \begin{smallmatrix} +.003 \\ -.000 \end{smallmatrix}$ HOLE
$\emptyset.380$ MMC
(−).005∅ ⊕ TOL AT MMC
$\emptyset.375$ GAGE PIN SIZE

DATUM D. TAPERED SPRING LOADED (OR
EXPANDING) PIN TO PICK UP ∅.500 TO ∅.502 HOLE)

RFS RELATED TO RFS DATUM FEATURE

When a pattern of holes is dimensioned relative to the location of another hole, this hole is identified as a datum and the hole pattern is located dimensionally with respect to it.

In the illustrated example, the four holes are related to the center datum hole. As the position of the center datum hole shifts, the position of the four hole pattern itself must follow as dictated by the function of the part.

In this instance, note that the four holes of the pattern are assigned an RFS position tolerance and that they are related to the datum hole D also on an RFS basis. Imagine that this part has precision requirements between the holes either to provide accurate relationship with a mating part or to maintain accuracy for a mating situation, such as a semicritical gear plate mounting.

Note that the .500 datum center hole is located from surface datums A, B, and C. It is given a position tolerance relative to these edges of ϕ .010 at MMC since the relationship to the edges can be on this basis with a rather lenient position tolerance. The orientation of the datum hole relative to the primary datum A is, however, to be maintained to a closer tolerance. Since datum D position and orientation are controlled on an MMC basis, the tolerances (position and perpendicularity) increase an amount equal to the produced size departure from MMC as shown in the illustration. Note, however, that the reference to datum D in the relationship of the four .380 holes is on an RFS basis. The four hole pattern, therefore, takes its positional reference from the exact center of the datum hole at whatever size it is produced (RFS) within .500 to .502.

Wherever the datum hole D position varies in the design considerations or in actual production, the four .380 hole pattern must follow. Note that the positional pattern dimensions originate at datum D to carry out this intent.

The four .380 holes are located by a position tolerance of ϕ .005 at RFS with respect to datum A (for attitude), datum D at RFS (for location), and datum B (for rotation). Reference to the explanation portion of the illustration (under Meaning) will assist in understanding the effect of the three datums and the importance of datum precedence. Although not illustrated, imagine that, as previously stated, the four .380 hole position tolerances have been determined to relate to a mating part or to maintain accuracy in a mating situation where other features or components must relate with precision regardless of the produced sizes (RFS) of the .380 holes (.380 to .383).

The four holes individually, with respect to their own true or exact positions, can vary in location up to ϕ .005. Under the RFS method, however, this tolerance applies to each hole regardless of the size to which it is produced. That is to say (as shown in the lower right corner of the illustration), the ϕ .005 position tolerance is the maximum allowable to each hole no matter to which size it is produced (.380 to .383). If, as was shown in previous illustrations, the MMC method had been applied, the position tolerance would increase to the extent of departure of the .380 holes from MMC. Not so, however, in this example since the RFS method has been invoked. The choice of proper approach, be it MMC or RFS, is of course decided by the design requirements. MMC methods are obviously recommended wherever possible due to the added functional (interchangeability) and tolerance advantages.

In this instance, it is seen that a more critical or precise relationship is maintained between the four holes and their datum. The design requirements based on part function determined this approach in our example.

AS DRAWN

MEANING

As may be seen in the previous examples in this section, functional gaging principles can be used to evaluate the datum hole location and relationships. However, the four hole (.380) pattern relative to datum D in this example cannot utilize such methods because it is an RFS application; open set-up principles would be necessary. The techniques of ''paper gaging and computation,'' described earlier in the text could, of course, be used as desired.

MMC RELATED TO MMC DATUM FEATURE

PROJECTED TOLERANCE ZONE*

This illustration shows a part with four tapped mounting holes. A cover plate mechanism is assumed as a mating part (not shown). Position tolerancing is used to assure assembly of the mating parts. The projected tolerance zone requirement is also added.

We wish to establish a relationship between the mounting surface, the mounting hole pattern, and the 1.506 counterbored seat diameter (identified as datum B). The mating part has a seat mechanism which must fit within this counterbore and attach and locate with screws to the four mounting holes on the flange surface.

The flange surface itself is established as the primary datum and is identified as datum A to ensure clarity of the hole pattern positional relationship with the top surface.

We calculate the position tolerances using the "fixed fastener" method as based on the .250–28 screw and the clearance hole in the mating part cover plate. Assuming that the cover plate has .256 clearance holes and .250 maximum thickness, we calculate the position tolerance and assign .003 to each part (.006/2 = .003).

The four .250–28 holes are thus designated as shown. The feature control symbol specifies "at true position within ⌀ .003 at MMC size of the holes with respect to datums A and B (at MMC) and to datum C."

Conventional position tolerancing discussed previously reveals that a position (MMC) application is affected by size departure of the involved features from MMC. However, in this instance, the peculiarity of thread assembly requires further consideration and may cause an exception to the usually inferred interpretation.

The centering effect of a screw as it tightens in the tapped hole tends to negate or diminish any *added* position tolerance based on greater clearance between the pitch diameters of the screw and tapped hole. The screw seeks center as, in tightening, the flank of the mating thread forms (screw and hole) come into contact or bottom out. Thus additional position tolerance relative to the tapped hole increase in pitch diameter size (departure from MMC) may not be fully realized. There may be some additional tolerance derived in actual assembly but it cannot be predicted.

The same reasoning as in the foregoing paragraph applies to attempts to assign position tolerance to countersinks. Position tolerance would be appropriate to the through hole, but, as a rule, is not practical for the countersink itself.

Added tolerance may be acquired, however, for the tapped hole pattern as a unit relative to datum B (counterbore) as it departs from MMC size (gets larger). Datum B is a straight sided feature, and size deviation from MMC will have the effect previously discussed in other examples.

---

*See also page 189 for further explanation of principle.

AS DRAWN

POSITIONAL TOLERANCE:
∅.003 AT FEATURE MMC
+∅.004 AT DATUM LMC LIMIT
∅.007 TOTAL TOL
POSSIBLE

SYMBOL MEANING

FOUR .250–28 HOLES MUST BE AT POSITION WITH RESPECT TO DATUM A & DATUM B (AT MMC) & DATUM C WITHIN ∅.003 AT MMC OF THE HOLES

AS THE DATUM FEATURE "B" SIZE INCREASES FROM MMC 1.504 TO 1.508 AN ADDITIONAL .004 IS ADDED TO THE POSITIONAL TOLERANCE OF THE PATTERN AS A GROUP.

MEANING

.250 MIN

AXIS PITCH
CYL THREAD

∅.003
POSN TOL
ZONE

.250 PROJ TOL

DATUM A

FUNCTIONAL GAGE (PRINCIPLES)

DATUM C SIM.
(ROTATION)

GO THREAD MEMBER GAGE PINS 4

∅X.XX PLUS POS TOL AT
MMC (.003) (4)

∅X.XX (GO MEMBER PIN (4)

∅1.504
(MMC) OF
SIM B DATUM (HOLE)
(LOCATION)

.250 (EQUAL TO PROJ TOL)

SIM DATUM A
(ORIENTATION)

Since the threaded holes are to attach a mating part, we see that in assembly the critical position location of these holes will actually concern the inserted screws or the actual projection from these holes which must accommodate the mating part. Therefore, the location of the threaded holes is controlled by a *projected* position tolerance zone to represent the projected screw locations. As is seen in the callout and sectioned projected tolerance zone view, the tolerance zone is cylindrical and extends above, and perpendicular to, the datum plane A. The drawing specification and view using the thick chain line clearly shows the direction and extent of the projected tolerance zone.

The height of the projected position tolerance zone is determined by the application. It may be established by the thickness of the mating part through which the screws are to extend, or it may be determined by the thread hole depth when a thin mating part is involved, or any desired height which fulfills the part design requirements may be selected. When the projected tolerance zone is intended, it should be specified as shown with the symbol immediately below the feature control frame (see illustration). This system is also used on parts in which pins, studs, or other features are to be inserted and the critical assembled location is *above* the surface of the part.

The representative production gage shown on the preceding page also illustrates this principle. The gage is made up of the main gage body containing the datum B hole pin, the plate which contacts the datum surface A, a spring plate to provide orientation to datum C, four GO thread members per standard tolerances and, if desired, torquing flats to represent appropriate screw tightening and setting. The plate contains four holes which are .003 (the positional tolerance) larger than the four GO thread pins which are inserted into the threaded holes of the part through the gage plate.

The relationship of the GO thread member projections and the holes in the gage plate simulate the functional assembly requirement of the two mating parts and the four screws.

Gage-maker's tolerances would, of course, also apply but are not shown.

MMC RELATED TO MMC DATUM FEATURE

The illustration on the opposite page shows a circular pattern of 8 holes located by position tolerancing with respect to the center hole.

The requirement states "the eight holes are to be located at position within ϕ .002 at .403 MMC size of the holes with respect to datum A and datum hole B (at 1.025 MMC)."

As the size of the holes increases in production from MMC .403 to the high limit of .405, .002 is added to the positional tolerance. As the datum hole B increases from MMC of 1.025 to 1.027, .002 is added to the hole pattern's positional tolerance as a group. Depending on the size increase of the holes, the total tolerance may be .006 instead of .002.

A production gage for this part is also shown. The pins in the pattern are spaced at 45° intervals on a 3.000 circle, with the pin size at .401 or at MMC size of the part hole .403 less than the positional tolerance. The center pin is 1.025 (MMC size of the datum hole).

Gage-maker's tolerances also apply.

AS DRAWN

⌀1.025 $+.002$ $-.000$

| ⊥ | ⌀.000 Ⓜ | A |

-B-

-A-

⌀3.000

8X 45°

8X ⌀.403 $+.002$ $-.000$

| ⊕ | ⌀.002 Ⓜ | A | B Ⓜ |

SYMBOL MEANING

POSITIONAL TOLERANCE:
⌀.002 AT HOLE MMC
+.002 AT HOLE HIGH LIMIT
⌀.004
+.002 AT DATUM HIGH LIMIT
⌀.006 TOTAL TOL POSSIBLE

EIGHT ⌀.403 HOLES TO BE AT TRUE POSITION WITHIN ⌀.002 (AT ⌀.403 MMC SIZE) WITH RE-SPECT TO DATUMS A AND B (AT ⌀1.025 MMC) SIZE

AS THE EIGHT HOLES INCREASE IN SIZE FROM MMC ⌀.403 TO HIGH LIMIT OF ⌀.405, .002 IS ADDED TO THE POSITIONAL TOLERANCE.

AS THE DATUM FEATURE B SIZE INCREASES FROM ⌀1.025 MMC TO ⌀1.027, .002 IS ADDED TO THE HOLE PATTERN POSITIONAL TOLERANCE AS A GROUP

GAGE

45° (8)

⌀3.000

⌀1.025 (MMC OF DATUM HOLE)

⌀.401 MMC OF HOLE .403 LESS POSITION TOL ⌀.002 (8) (LOCATION)

SIM DATUM A (ORIENTATION)

This example shows the use of positional tolerancing on two locating pins on a part. The two pins are to take reference from the center hole which accommodates a shaft mechanism of the mating part (not shown). The center hole is selected as the datum reference. This example reminds us that positional tolerancing is not restricted to holes. It can also be applied to pins or to any feature on which a center (axis) is the basis for location.

The meaning of the position requirement is, "the two pins are to be located at true position within ⌀ .002 tolerance zone at .125 MMC size of the pins with respect to datums A, B (at MMC .500), and C."

Note that MMC of the pins is the high limit of size. As the two pins reduce in size from MMC of .125 to the low limit (LMC) of .122, .003 is added to the positional tolerance. As the datum feature B (the center hole) increases from MMC of .500 to .502 high limit, .002 is added to the two pins' positional tolerance as a group.

Depending on the production sizes of the bosses and datum hole, the total positional tolerance on the boss location may vary from ⌀ .002 to .007.

A representative gage to check the position location and the relationship with the datum hole is also shown. Note that the holes to check the position location of the pins are to the high limit of MMC of the pins *plus* the position diameter tolerance. The datum hole pin is to the MMC size of the datum hole of .500.

Gage-maker's tolerances also apply.

AS DRAWN

⌀.500 $^{+.002}_{-.000}$

| ⊥ | ⌀.000 Ⓜ | A |

-B-

-A-

.130 ±.005

3.14 ±.02

1.57 ±.01

-C-

.500

1.000

1.130

2.260

2.25 ±.02

1.125 ±.010

2x ⌀.125 $^{+.000}_{-.003}$

| ⊕ | ⌀.002 Ⓜ | A | B Ⓜ | C |

POSITIONAL TOLERANCE:
⌀.002 AT PIN MMC SIZE
+.003 AT PIN LOW LIMIT
⌀.005
+.002 AT DATUM HIGH LIMIT
⌀.007 TOTAL TOL POSSIBLE

SYMBOL MEANING
TWO PINS TO BE AT TRUE POSITION WITHIN
⌀.002 TOL ZONE AT MMC SIZE WITH RESPECT
TO DATUMS A, B (.500 MMC), AND C.

AS THE TWO PINS REDUCE IN SIZE FROM MMC
OF ⌀.125 TO LOW LIMIT (LMC) OF ⌀.122, .003 IS
ADDED TO THE POSITIONAL TOLERANCE.

AS THE DATUM FEATURE "B" (THE CENTER
HOLE) INCREASES FROM MMC OF ⌀.500 TO
⌀.502 HIGH LIMIT, .002 IS ADDED TO THE TWO
PINS POSITIONAL TOLERANCE AS A GROUP.

GAGE

SIM
DATUM C (ROTATION)

.500

1.000

⌀.127 HOLE (2)
(MMC OF
PIN .125
PLUS
POSITION
TOL ⌀.002)

1.130

2.260

⌀.500 MMC
OF DATUM B
HOLE
(LOCATION)

SIM
DATUM A (ORIENTATION)

POSITION TOLERANCE—LEAST MATERIAL CONDITION (LMC)

Occasionally we need a method to control a situation which is essentially the reverse of the usual position relationship; that is, the stated position tolerance applies at the *least material condition,* LMC, of the feature or datum, instead of at MMC, and increases as the feature or datum *departs from* the least material condition.

Definition. Least material condition (LMC) is the condition in which a feature of size contains the least amount of material within the stated limits of size: for example, maximum hole diameter, minimum shaft diameter. Least material condition is the condition opposite to MMC. For example, a shaft is at least material condition when it is at its *low* limit of size and a hole is at least material condition when it is at its *high* limit of size.

This method is applicable to special design requirements that will not permit MMC or that do not warrant the exacting requirements of RFS. It can be used to maintain critical wall thickness or critical center locations of features for which accuracy of location can be relaxed (position tolerance increased) when the feature leaves least material condition and approaches MMC. The amount of increase of positional tolerance permissible is equal to the feature size departure from least material condition.

The term "least material condition" and the abbreviation LMC have been used instead of "minimum material condition" (which is synonymous) to avoid confusion, since the abbreviation would be the same as that for maximum material condition. The symbol modifier Ⓛ is used to indicate the LMC requirement applicable to feature or datum.

Although the use of LMC does impose exacting requirements on both manufacturing and inspection, it does permit additional tolerances.

Whenever least material condition (LMC) or Ⓛ is specified on a drawing, the position tolerance applies only when the feature is produced at its LMC size. See Fig. 1.

Additional positional tolerance is permissible but is dependent on, and equal to, the difference between the actually produced feature size (within its size tolerance) and LMC. See Fig. 2.

See Fig. 3 for an additional example of LMC applied to non-cylindrical features. As shown under MEANING, the tolerance zone increases as the feature size departs from LMC toward MMC.

POSITION TOLERANCE—LEAST MATERIAL CONDITION (LMC)

EXAMPLE

MEANING

FIGURE 1

- Ø.303 LMC (HIGH LIMIT) HOLE
- Ø.005 POSITION TOL ZONE
- POSITION AT MAXIMUM OF POSITION TOL ZONE

FIGURE 2

- Ø.300 LOW LIMIT (MMC) HOLE
- Ø.008 POSITION TOL ZONE (.005 POSITION TOL +.003 SIZE TOL)
- POSITION AT MAXIMUM OF POSITION TOL ZONE

EXAMPLE

FIGURE 3

MEANING

LMC RELATED TO RFS DATUM FEATURE

When a pattern of holes is dimensioned relative to the location of another hole, this hole is identified as a datum and the hole pattern is located dimensionally with respect to it. In this case, as contrasted with some earlier examples of similar parts using MMC, the holes are controlled on an LMC basis with a more critical relationship (RFS) to the datum center hole. In a situation of this kind the effect of the MMC principle could be detrimental to the necessary precision of the design requirement, so the design specifies the reverse principle, LMC. Such applications as gear centers, or other critical component interfaces, could require such considerations. LMC provides a compensating effect which states in essence that the hole location criterion is at its LMC size ("sloppiest") and can permit an increase in position tolerance as it (the hole) reduces in size (toward MMC) and "improves" the component/fit relationship. This also provides a dynamic alternative to specifying RFS to the four hole location.

In the example, the four holes are related to the center datum hole. As the position of the center datum hole shifts, the position of the four hole pattern itself must follow as dictated by the function of the part. Imagine that this part has a mating part with a shaft and four pins which must assemble with the five holes in the illustrated part.

Note that the .500 datum center hole is also located from datums (surfaces) A, B, and C. It is given a position tolerance of ⌀ .010 at MMC and a refined orientation, or perpendicularity, tolerance of ⌀ .003 at MMC. This properly controls the center hole relationship to the edge surfaces with a rather lenient position tolerance while the orientation to the primary datum A is maintained to closer tolerance. The center hole is identified as datum D so that the four .380 holes can be located with respect to it. Under MMC principles, the position and perpendicularity tolerances increase an amount equal to the hole size departure from MMC as shown in the illustration.

Wherever the datum hole D position varies in the design considerations, or in actual production, the four .380 hole pattern must follow. Note that the positional pattern dimensions originate at datum D to carry out this intent.

The four .380 holes are located by a position tolerance of ⌀ .005 at LMC with respect to datum A (for orientation), datum D regardless of feature size (RFS) (for location), and datum B (for rotation). Reference to the lower portion of the illustration under Meaning will assist in the understanding of the effect of the three datums and the importance of datum precedence. Although not illustrated, imagine the four .380 hole position tolerances have been calculated relative to a mating part using the "fixed fastener" method previously explained. However, the LMC principle may require added considerations and special calculations due to its unique effect and as based on the design intent.

The four holes individually with respect to their own true positions can vary in location up to ⌀ .005 at LMC. While at any size in a departure from LMC a hole can vary an additional amount (enlarges positional tolerance) equal to that departure. For example (as shown), if the hole size is produced at .382, the position tolerance becomes ⌀ .006, at .381 it becomes ⌀ .007, etc., up to ⌀ .008 if the hole is produced at .380 MMC.

The datum feature D is referenced on an RFS basis. This neutralizes the effect of size departure from MMC of the datum feature to the related hole pattern movement. It must be located from the axis of datum D and cannot move with respect to that axis. This could be in keeping with the precision of the part interface, the LMC requirement, and also could be to restrict the pattern movement as a concern for wall thickness between the outer hole surface limits relative to the outside surfaces. Where deemed necessary to design intent, LMC could also be applied to datum D in its established relationship to the small hole pattern. In which case, the four hole pattern, as a group, could shift an amount equal to the amount of such departure from LMC of the datum hole D.

Functional gages and comparable techniques could be applied in the verification of the $\emptyset$ .500 datum hole requirements as earlier illustrated on a similar part. However, functional gaging, of the physical variety, could *not* be performed on the four hole LMC requirement. It could be done with optical methods, graphic analysis, mathematically, etc., where the functional principles of LMC can be simulated as a gaging operation.

The advantages of LMC are (1) that it provides an often needed tool and (2) that it also provides an alternative other than RFS, where the MMC method is not desirable.

See also earlier examples in this section to determine the best choice of application on such pilot hole applications.

AS DRAWN

MEANING

∅ SIZE	∅ TOL
.500 (MMC)	.010
.5005	.0105
.501	.011
.5015	.0115
.502 (LMC)	.012

DATUM HOLE POSITION TOL ZONE ∅
.010 (AT MMC)
.012 (AT LMC)

DATUM PERP. TOL ZONE ∅
.003 (AT MMC)
.005 (AT LMC)

∅ SIZE	∅ TOL
.500 (MMC)	.003
.5005	.0035
.501	.004
.5015	.0045
.502 (LMC)	.005

POSITION TOL ZONE ∅ (4)
.005 (AT LMC)
.008 (AT MMC)

∅ SIZE	⊕ TOL
.383 (LMC)	.005
.3825	.0055
.382	.006
.3815	.0065
.381	.007
.3805	.0075
.380 (MMC)	.008

POSITION
OF
NONCYLINDRICAL
FEATURES

ASSEMBLY OF TWO PARTS WITH NON-CYLINDRICAL
FEATURES (see following pages)

Position tolerance relationships are more often associated with round holes or features and establish a cylindrical tolerance zone around theoretically exact axes. The cylindrical tolerance zone is not applicable to slots, dial markings, tabs, etc., for which non-cumulative tolerance and MMC aspects of positional tolerancing may also be desired. Many such parts could be considered "symmetrical" and would possibly have been stated as a "symmetry" requirement under former standards.

Such features may be allowed to vary with respect to a center plane rather than an axis. The position tolerance zone is a total wide zone with one-half the total tolerance assigned to each side of the center plane.

In this example, as based upon an assembly relationship shown on the preceding page, two mating parts illustrate the calculations and relationships. The top part could be either a thin metal part or a type of drive shaft with three tab projections. The mating part below might be a sleeve or collar which must fit the upper part. To simplify initial explanation, side views and primary (orientation) datums are not shown on either part.

Both parts have corresponding datum reference diameters which are related, in turn, to the controlled features of each individual part. The datums are identified by the letter A in the datum feature symbol. The feature control frame for the top part (Example 1) reads, "these features (3 tabs) must be at true position within .006 total wide zone with the feature at MMC size and with respect to datum A at MMC." Although the symbol used is the same as that for cylindrical zones, there is no confusion, since the drawing always clearly shows the feature being dimensioned and the ⌀ symbol is not used, thus designating the tolerance zone as a noncylindrical width zone.

The feature control frame for the bottom part (Example 2) reads, "these features (3 slots) must be at true position within .006 total wide zone with the features at MMC size and with respect to datum A at MMC."

Note that the tolerance zones are *not* cylindrical but are total widths (parallelepiped) equally disposed about, and parallel to, the center plane as established by the 120° basic angles and extending the full depth and length of the produced feature.

When designated as shown, the width of the tolerance zone is always total and is equally disposed on either side of the true position center plane. In this case, the total wide zone is .006, with .003 on each side of the basic center plane.

The calculations of the positional tolerance zones for mating parts of this type are shown at the upper right. They are based on the same reasoning previously discussed for "fixed fasteners" using cylindrical features. The tolerance zones in this case are, however, not cylindrical but total width.

As in any positional calculation, the MMC sizes of the two corresponding mating part features are used to determine their individual positional tolerances. The MMC width of the tab, .250, is subtracted from the MMC width of the slot, .262, giving a combined clearance of .012. This is divided by the fixed factor 2 to give the total tolerance zone for each mating part feature at MMC. As previously discussed on round feature positional tolerance calculations, the total combined tolerance (in this case .012) may be divided as desired in other combinations (e.g., .008 and .004, .009 and .003, etc.). See page 161 for further detail on calculation methods.

NONCYLINDRICAL MATING PART FEATURES

EXAMPLE 1

MEANING

PART ACCEPTANCE TOL

TAB WIDTHS AT MMC
.250 (HIGH LIMIT)
POSITIONAL TOL = .006 TOTAL
WIDE ZONE
TAB WIDTHS AT LMC (LOW
LIMIT) .245
POSITIONAL TOL = .011 TOTAL
WIDE ZONE

3X .250 $^{+.000}_{-.005}$

⊕ | .006 Ⓜ | A Ⓜ

Ø1.500
+.000
−.003

-A-

GAGE

120° (3)

.256 (3)

1.500
(MMC DATUM A)

.006 TOTAL
WIDE TOL ZONE
(AT TAB MMC)

.003
.003

.250 MMC
PART
+.006 POSITIONAL
TOL
.256 GAGE SLOT

CALCULATIONS

MMC WIDTH OF SLOT = .262
MMC WIDTH OF TAB = .250(−)
2 | .012

TOTAL WIDE TOL
ZONE = .006
FOR SLOTS & TABS
AT MMC
(.003 EACH SIDE OF BASIC
TRUE POSITION CENTER PLANE)

EXAMPLE 2

MEANING

PART ACCEPTANCE TOL

SLOT WIDTHS AT MMC
.262 (LOW LIMIT)
POSITIONAL TOL = .006 TOTAL
WIDE ZONE
SLOT WIDTHS AT LMC (HIGH
LIMIT) .267
POSITIONAL TOL = .011 TOTAL
WIDE ZONE

3X 120°

3X .262 $^{+.005}_{-.000}$

⊕ | .006 Ⓜ | A Ⓜ

Ø1.505
+.003
−.000

-A-   1.505
(MMC DATUM A)

GAGE

120° (3)

.256 (3)

.006 TOTAL
WIDE TOL ZONE
(AT SLOT MMC)

.003
.003

.262 MMC PART
−.006 POSITIONAL TOL
.256 GAGE TAB

In Example 1, the notation "Part Acceptance Tolerance" indicates that the total positional tolerance zone increases from .006 to .011 as the actually produced tab width reduces from MMC of .250 to .245.

The same is true for Example 2. The slot width positional tolerance increases from .006 to .011 as the slot is produced to the high limit size of .267. Simulated gages are also shown.

Note that the .250 MMC tab width in Example 1 is accommodated by a .256 gage slot determined by adding the .006 positional tolerance to the MMC size of the part tab. The gage for Example 2 shows the reverse, with the positional tolerance of .006 being subtracted from the .262 MMC size of the part slot to establish the gage tab size of .256.

Functionally gaging these parts will permit additional positional tolerance for the tabs and slots as a group as the datum feature sizes depart from MMC. However, because of the unique geometry, the actual amount of this effect (due to the datum diameter relationship to the slot and tab flats) may not be conveniently predicted or calculated.

MMC WITH RESPECT TO A LINE (DIAL) AND RELATED DATUM FEATURE

Positional tolerancing may also be applied effectively to parts which do not mate. Since position tolerancing is noncumulative, it is ideally suited to dial markings, etc.

The example illustrates a dial with graduated markings which have critical angular location with respect to one another, and with the center hole identified as the datum reference. The length of the markings are relatively unimportant. In cases such as this, it is usually important that the angular distances between the markings not accumulate tolerance. The use of basic angle locations with position tolerance zones straddling these locations is ideal.

Referring to Meaning, note that the tolerance zones for the groove center lines are parallel to the 15° basic center planes which are taken from the actual center of the .190 datum hole A. The tolerance zones are total *wide* tolerance zones with half the zone on each side of these basic center planes. For each marking, the tolerance zone is .010 total when the marking grooves are at MMC size of .030. The tolerance zones can increase to as much as .015 if the dial markings are produced to their least material condition (high limit) of .035.

The center hole datum feature is specified at RFS with the RFS symbol included with the datum reference letter A. The RFS datum is used to ensure that the size variation of the center datum hole will not affect the position pattern location with respect to the center hole axis.

USE OF "LEAST MATERIAL CONDITION" APPLICATION*

When greatest accuracy of the slot angular position is required while the slot marking is at its largest width .035 (high limit size), the principles of "least material condition" (LMC) may be desired.

---

*See pages 144–149.

EXAMPLE

7X .035 $^{+.000}_{-.005}$ WIDE × .005/.010 DEEP

⊕ | .010 ⓛ | A Ⓢ

MEANING

These markings are to be located at position within .010 total wide zone at marking LMC size of .035 with respect to datum A (RFS). As the marking groove gets smaller, or departs from LMC (approaches MMC), the position tolerance is increased an equal amount; e.g., at the marking groove size of .030 width, the position tolerance is .015 total wide zone.

AS DRAWN

6X 15°

7X .030 $^{+.005}_{-.000}$
▽ .005 −.010

⊕ | .010 Ⓜ | A Ⓢ

SYMBOL MEANING

⊕ | .010 Ⓜ | A Ⓢ → WITH RESPECT TO DATUM A, (RFS)

WITHIN .010 TOTAL WIDE ZONE AT MARKING MMC SIZE

THESE MARKINGS TO BE LOCATED AT TRUE POSITION

7X .625 ±.010

Ø1.500 ±.005

Ø.190 $^{+.005}_{-.000}$

−A−

MEANING

MARKING CENTERLINE TOTAL WIDE TOL ZONE AT MMC

.010

.005

15° (6)

MAY INCREASE TO .015 MARKING LMC (HIGH LIMIT) SIZE

ACTUAL CENTER OF .190 $^{+.005}_{-.000}$ DATUM A (RFS)

POSITION (MMC) WITH RESPECT TO CENTER PLANES (MATING PARTS)

This illustration presents a unique mating part relationship. Position tolerancing principles are extended to utilize a combination of total wide position requirements. The horizontal slots of the upper part are to mate with screw clearance holes of the lower part.

The illustration clearly shows that assembly requirements demand closer control of the hole location in the vertical direction than in the horizontal direction since the slots can compensate for considerable horizontal variation. Therefore the conventional cylindrical zone position method does not seem appropriate but total wide zone methods appear to satisfy the requirement. Using this approach, we can specify separate positional tolerance zones of different values for vertical and horizontal displacements of the slot and holes.

In Example 1, it is quite apparent from the elongated slots that a width type positional tolerancing rather than diametral positional tolerancing is desired. However, in Example 1 as well as in Example 2, the absence of the diameter symbol $\emptyset$ and the direction of the arrows is the key indicator that the total wide tolerance zone is to apply. The vertical and horizontal extension lines from the features indicate the directions of the tolerance zones, and the datums and tolerance of the feature control frame state the orientation and total width of the tolerance zone. Note that the tolerance zone consists of the total width equally disposed over the center planes; it extends the full depth of the feature perpendicular to the primary datum.

One may ask why the positional tolerancing system was used here, since the resulting tolerance zone seems to be the same as that which would have been obtained from coordinate dimensioning and tolerancing. The answer is that the positional MMC system recognizes that the *size* to which these features are produced affects their position relationships. As the feature sizes depart from MMC, that is, as the slots or holes get larger in their size tolerance range, additional positional tolerance is permitted. For instance, in Example 1, the .012 total wide tolerance zone on the slots could be increased to as much as .017, and the .012 and .030 total wide zones on the holes in Example 2 could be increased to as much as .017 and .035, respectively, depending on the sizes to which these features are actually produced.

Functional gaging techniques could be used on this part.

See also Extended Use of Position Tolerancing for similar applications.

EXAMPLE 1

4X .262 +.005 / −.000

⊕ | .012 Ⓜ | A | B

.500

1.000

SYMBOL MEANING

THESE SLOTS TO BE LOCATED AT TRUE
POSITION WITHIN .012 TOTAL WIDE
ZONE AT SLOT MMC SIZE OF .262
WITH ORIENTATION FROM DATUM
PLANES A & B.

−B−

−A−

2.880 ±.020

2X .55 ±.01

1.000

.006

.012

PER COORDINATE
DIMS & TOL

EXAMPLE 2

−C−

3.245

4X
⊕ | .030 Ⓜ | A | B | C

.500

4X
⊕ | .012 Ⓜ | A | B | C

1.000

−B−

.250

4X ∅.262 +.005 / −.000

−A−

SYMBOL MEANING

THESE HOLES TO BE
LOCATED AT TRUE POSITION
WITHIN .030 (HORIZONTAL)
& .012 (VERTICAL) TOTAL
WIDE ZONE AT MMC HOLE
SIZE .262 WITH ORIENTATION
FROM DATUMS A, B, & C.

.012   1.000
.006

.030
.015

3.245

.250

MMC—NONCYLINDRICAL MATING PART FEATURES

This illustration shows a pair of mating parts involving noncylindrical features. Part 1 is to fit within the opening of part 2.

Part 1 has a width of $1.000 \, {}^{+.000}_{-.006}$ which is to fit within the $1.005 \, {}^{+.003}_{-.000}$ opening width on part 2. Simultaneously, the $.500 \, {}^{+.004}_{-.000}$ slot on part 1 is to fit onto the $.495 \, {}^{+.000}_{-.003}$ projection on part 2.

The $.500 \, {}^{+.004}_{-.000}$ slot on part 1 has a position feature control symbol which states, "this feature is to be at true position within .005 at MMC size of the feature with respect to datum A at MMC size." The width of the part is established as datum A.

Part 2 has an identical position feature control frame on the $.495 \, {}^{+.000}_{-.003}$ dimension, and the $1.005 \, {}^{+.003}_{-.000}$ opening is established as datum A.

Figure 1(a) shows the relationship of these two parts as they would appear if both parts were produced perfectly at the feature MMC sizes. Note the common center or median planes established on both parts. The parts are assembled in Fig. 1(b).

Figure 2(a) illustrates the slot feature on part 1 offset the maximum permissible amount (.0025) at the extreme of the .005 total tolerance zone when the part is at MMC size. Also, the mating projection of part 2(b) is shown offset in the opposite direction the maximum permissible amount (.0025) at the extreme of the .005 total tolerance zone when the part is at MMC size.

Figure 2(b) shows the assembly of the two parts. They still assemble satisfactorily. Figure 2 also emphasizes that the .005 total tolerance zone, as stated in the symbol boxes on parts 1 and 2, applies at the MMC size of the features and is the maximum tolerance permissible under this condition.

Figure 3 illustrates the increase in the permissible total position tolerance zone as the feature sizes *depart* from MMC to the opposite extreme of LEAST MATERIAL CONDITION. For part 1 (Fig. 3a), with the slot at its *high* limit size of .504 and the datum width at its *low* limit of .994, the permissible position tolerance zone becomes .015 total or a .0075 offset off the median plane of the slot with respect to the datum median plane.

For part 2 in Fig. 3(b), with the projection at its low limit of .492 and the datum opening width at its high limit of 1.008, the position tolerance zone becomes .011 total or a .0055 offset off the median plane of the projection with respect to the datum median plane.

Figure 3(b) shows the assembly of the two parts under these conditions. They still assemble satisfactorily with considerably more clearance as a result of the feature size variation to size limits opposite MMC, or their LEAST MATERIAL CONDITION.

From these illustrations it is evident that positional MMC applications permit greater tolerance and ensure a satisfactory fit of mating parts. For example, the possible tolerance on part 1 has been increased from .005 to .015 and on part 2 from .005 to .011. The actual tolerance to be realized is, of course, dependent upon the sizes to which the concerned features are actually produced.

MAXIMUM MATERIAL CONDITION (MMC)
EXAMPLES

PART 1

PART 2

PERFECT POSITION AT MMC
MEANING

FIGURE 1   (a)   (b)

POSITION TOLERANCE ZONES AT MMC

FIGURE 2   (a)   (b)

POSITION TOLERANCE ZONES AT LEAST
MATERIAL CONDITION

FIGURE 3   (a)   (b)

NONCYLINDRICAL MATING PART FEATURES

MMC CALCULATIONS TO DETERMINE TOLERANCE

In this example we present the calculations required to determine the positional tolerance for the mating parts shown in the previous example.

Since one part is to fit within the other, the first step is to determine the clearance of the features and which feature is to receive the position tolerance. In this case, it seems more functional to control the position of the slot in part 1 and the position of the projection in part 2. The clearance of the two mating part features is to be .005 minimum. The projection on part 2 is .495 and the slot on part 1 is larger at .500. These are MMC sizes, or the largest projection possible on part 2 and the smallest slot possible on part 1.

The width features on both parts are given .005 clearance at MMC size of the features and are selected as the datum features for each part.

Under the subheading, POSITION TOLERANCE CALCULATIONS, the .495 MMC size of the projection on part 2 is subtracted from the .500 MMC size of the slot on part 1. This results in a difference of .005. Next the 1.000 MMC datum projection feature of part 1 is subtracted from the 1.005 MMC datum slot of part 2, resulting in a difference of .005.

The .005 result of the first calculation and the .005 result of the second calculation are added to give the .010 total combined positional tolerance for both parts and their interrelated features. This total tolerance is then divided to establish the required position tolerance on *each* individual part. How we allocate the total tolerance is optional, so long as it totals the calculated combined tolerance, in this case .010.

For the purposes of this example, the .010 total tolerance was divided evenly, with .005 selected as the position tolerance for both the .500 slot on part 1 and the .495 projection on part 2. These two figures, .005 plus .005, total .010 and comply with the .010 allowable total combined positional tolerance calculated.

Once the position tolerance is established for both mating part features based on their relationship to each other and to common datum axes, possible extra position tolerance for each part may be determined as shown in the lower half of the figure.

To do these calculations, we must first determine the relationship of one mating part feature to another and, then, we must consider each of these part features individually with respect to the size variations which could occur within their size tolerances. As has been shown, the size of features affects their positional tolerances, and it is this fact that makes positional tolerancing advantageous, since it permits economical production with greater tolerances and ensures assembly of the mating parts.

On part 1, the permissible tolerance may be increased from .005 to .015 and on part 2 the permissible tolerance may be increased from .005 to .011. The *actual* tolerance permissible in each case is, of course, dependent on the *actual* sizes of the features as produced.

This method of calculating position tolerances assumes the possibility of zero clearance — zero interference fits of mating part features if all features are at extreme tolerance limits. It also assumes parallel orientation or permissible float of one part to the other at assembly. Additional compensation of the calculated tolerance values should be considered as necessary for any particular application or where additional datum orientation may restrict this float.

EXAMPLES

PART 1

PART 2

MATING PARTS

POSITION TOLERANCE CALCULATIONS

MMC SIZE SLOT (PART #1)   = .500
MMC SIZE PROJECTION (PART #2)   = (−).495
                                                              .005 − − − − − .005
MMC SIZE DATUM SLOT (PART #2)   = 1.005   (+).005
MMC SIZE DATUM PROJECTION   (−)1.000   .010 =
                         (PART #1)   .005 −

TOTAL TOLERANCES
TO BE DIVIDED AS DESIRED
TO ESTABLISH REQUIRED
POSITION TOLERANCE
ON EACH INDIVIDUAL
PART. *CAN BE ANY
COMBINATION WHICH
TOTALS .010".*

e.g. SELECTED .005 FOR PART #1 & .005 FOR PART #2

$$T = \frac{(S - P) + (D_1 - D_2)}{2} \qquad 2T = T_1 + T_2$$

S = SLOT   PART 1
P = PROJ   PART 2
D_1 = DATUM SLOT PART 2
D_2 = DATUM PROJ PART 1

EXTRA TOLERANCE FOR EACH PART

PERMISSIBLE SLOT POSITION TOL AS FEATURE SIZES DEPART FROM MMC:

STATED POSITION TOL WITH SLOT AT .500 MMC   = .005
PLUS TOTAL .500 SLOT SIZE TOL   + .004

POSITION TOL WITH DATUM WIDTH AT 1.000 MMC   = .009
PLUS TOTAL 1.000 DATUM WIDTH SIZE TOL   + .006

TOTAL POSITION TOL WITH SLOT & DATUM WIDTH AT   = .015
LEAST MAT'L CONDITION (LARGEST SLOT, SMALLEST
DATUM WIDTH)

PART 1

PERMISSIBLE PROJECTION POSITION TOL AS FEATURE SIZES
                                             DEPART FROM MMC:
STATED POSITION TOL WITH PROJECTION AT .495 MMC   = .005
PLUS TOTAL .495 PROJECTION SIZE TOL   + .003

POSITION TOL WITH DATUM OPENING AT 1.005 MMC   = .008
PLUS TOTAL 1.005 DATUM SLOT SIZE TOL   + .003

TOTAL POSITION TOL WITH PROJECTION & DATUM   = .011
OPENING AT LEAST MAT'L CONDITION (SMALLEST PROJ.,
LARGEST DATUM OPENING)

PART 2

NONCYLINDRICAL MATING PART FEATURE GAGES

Functional gages may be utilized on noncylindrical parts when position tolerancing is used.

Part 1 shown at the right is the same part used in the preceding examples. Immediately below part 1 is a representative functional gage used to check position on this part. Gage-maker's tolerances would, of course, be included in the actual construction of this gage.

From the previous explanations and this illustration, we see that as the part feature sizes depart from MMC within their size tolerances, they become equally acceptable to the gage while permitting greater position tolerance. The gage essentially simulates the mating part situation, and therefore a part which passes the gage will assemble with its mating part.

Note, however, that the high and low limits of size are determined by other gages or measurements. The illustrated gage checks only the position requirement.

Part 2 is the mating part used in the previous examples. Below it is a representative functional gage used to check the position on this part. Gage-maker's tolerances would, of course, also be included in the actual construction of this gage.

As in the mating part, above, we see that as the part feature sizes of part 2 depart from MMC within their size tolerances, they become equally acceptable to the gage while permitting greater position tolerance. The gage essentially simulates the mating part situation, and therefore a part which passes the gage will assemble with its mating part.

As in part 1, the high and low limits of size are determined by other gages or measurements. The illustrated gage checks the position requirement only.

PART 1

1.000 +.000 / −.006 -A-

.500 +.004 / −.000

⊕ | .005 Ⓜ | A Ⓜ

MMC FUNCTIONAL GAGE

1.000*

.500

.2475

.495*

*GAGE MAKERS TOL AS REQUIRED

GAGE CALCULATIONS

MMC SIZE PART SLOT = .500
  MINUS POSN
    TOL = (−).005
  GAGE SIZE = .495

MMC SIZE OF PART
DATUM WIDTH (1.000)
ESTABLISHES GAGE
WIDTH SIZE = 1.000

PART 2

1.005 +.003 / −.000 -A-

.495 +.000 / −.003

⊕ | .005 Ⓜ | A Ⓜ

MMC FUNCTIONAL GAGE

1.005*

.250

.500*

.5025

*GAGE MAKERS TOL AS REQUIRED

GAGE CALCULATIONS

MMC SIZE PART PROJ. = .495
  PLUS POSN TOL = (+).005
  GAGE SIZE = .500

MMC SIZE OF PART
DATUM WIDTH (1.005)
ESTABLISHES GAGE
WIDTH SIZE = 1.005

MMC–NONCYLINDRICAL FEATURES

The example below illustrates control of a flat feature in a position relationship to a cylindrical datum. This situation is typical of many parts for which position functional techniques will assure agreement with design intent and also permit functional or receiver gaging.

The part is shown with a representative functional gage. Note that the possible position tolerance increases from .010 to .016 maximum as the feature sizes depart from MMC toward the opposite or least material limits. The actual position tolerance depends on the sizes to which the features are produced with the amount of increase equal to the departure from MMC size.

EXAMPLE

GAGE

GAGE CALCULATIONS	
MMC SIZE PART	= .073
PLUS POSN TOL	= (+) .010
GAGE	.083*
MMC SIZE PART DATUM DIA ESTABLISHES	= .122
GAGE DIA. SIZE	= .122*

*GAGE MAKERS TOL AS REQUIRED

RFS–NONCYLINDRICAL FEATURES

Noncylindrical feature application may require RFS precision. In the case of this example, the part is also a symmetrical one where the features must be controlled with respect to their location to one another.

The example shows a part using position tolerancing on the slot; the part width is established as the datum feature. The requirement is to relate the slot location to the outside width of the part. To simplify explanation, size dimensions and tolerances of the slot and the datum width have been omitted.

In order to establish some basis for the relationship of the slot feature to the datum feature, we must have a common reference plane. This is the center, or median plane, of the datum feature, and thus it is the datum center plane and also the true position center plane. We also determine the resulting center plane of the toleranced feature and then compare these two center planes. The results of this comparison show whether or not the tolerance is met.

The interpretation shows the relationship and establishment of the position tolerance zone with respect to the datum center plane. Note that the center plane, or median plane, of datum A establishes the middle (true position) of the .005 tolerance zone.

The median plane of the produced slot, regardless of the slot size (RFS), must lie between these two planes, which are .005 apart and equidistant (or .0025) from the center plane of the datum and true position, regardless of datum size.

Note that the center plane of the slot, as determined by the actual slot, may fall at an angle or anywhere within the tolerance zone.

EXAMPLE

MEANING

RFS–NONCYLINDRICAL FEATURES

In this example, a part similar to that in the previous illustration has been used. Size dimensions and tolerances have also been added to the slot and datum features and an added datum has been specified. This example, as the previous one, is also an RFS application.

The meaning below the example in Fig. 1a illustrates the establishment of the .005 tolerance zone about the center plane of the datum feature. Note that the tolerance zone remains the same, .005, when the datum feature size and the slot size vary anywhere in their size tolerance range. See Fig. 1b.

In either case, and also at any place between these sizes, the center plane is established by the midpoint of the datum size but the .005 tolerance zone remains the maximum permissible tolerance.

The total tolerance zone for the slot is therefore .005 regardless of the size of the datum feature or the slot.

In Figures 2(a) and 2(b) we show the manner in which the center plane of the slot, regardless of its size, may vary within the position tolerance zone. In 2(a) the slot is shown at .500 and in 2(b) at .504. To be acceptable, the center planes of either of these extremes of the slot size must fall within the .005 tolerance zone. Note that the slot center planes in either example fall within the tolerance zone so these parts are acceptable.

Figure 3 shows that the slot center plane may fall anywhere within this tolerance zone.

In Figure 4 we show one method of verifying the position tolerance on this part in order to clarify how theoretical center planes presented on drawings are converted to actual measurements. The slot has been offset to the limit of the .005 tolerance zone. A .500 wide slot and a 1.000 wide datum feature have been assumed for this example. The part is rested against datum plane B (primary) and on one side of the 1.000 datum feature A, or its width. A maximum measurement is taken from the measuring surface to the top underside of the slot.

The part is then rotated 180°, or turned over, to the opposite side of the 1.000 datum feature A and against datum plane B (primary) and a similar maximum measurement is taken. The difference between these two measurements, as may be seen in the illustration, actually determines the total variation of the slot center plane with respect to the datum center plane and the tolerance zone.

A preferred (and actually more correct) method, although usually requiring more set-up equipment, is to establish the datum center plane simultaneously from the part width. See the DATUM section for more detail.

Comparison between the resulting value and the tolerance requirement determines the acceptance of the slot position.

As seen in this example, the resulting .005 total variation of the measurements is equal to the .005 required tolerance. Therefore, the slot is positional, or central, to datum A within the required tolerance and is also square (perpendicular) to the primary datum B.

EXAMPLE

MEANING

TOTAL TOL ZONE FOR SLOT IS .005 REGARDLESS OF SIZE OF DATUM OR SLOT

FIGURE 1(a)   **FIGURE 1(b)**

CENTER PLANE OF SLOT, REGARDLESS OF SLOT SIZE, MUST FALL WITHIN .005 TOTAL TOL ZONE

FIGURE 2(a)   **FIGURE 2(b)**

FIGURE 3

GAGING

FIGURE 4
*PERPENDICULAR TO DATUM
PLANE B

DIFFERENCE = .005

TOTAL VARIATION. THEREFORE SLOT IS SYMMETRICAL
OR CENTRAL TO DATUM A WITHIN THE REQUIRED TOLERANCE.

POSITION
OF
COAXIAL
FEATURES

COAXIAL FEATURES—SELECTION OF PROPER CONTROL

There are three methods of controlling interrelated coaxial features:

1. RUNOUT TOLERANCE (circular or total) (RFS)

2. POSITION TOLERANCE   (MMC)

3. CONCENTRICITY TOLERANCE   (RFS)

Any of the above methods provides effective control. However it is important to select the *most appropriate* one to both meet the design requirements and provide the most economical manufacturing conditions. (See also details of preceding sections).

Below are recommendations to assist in selecting the proper control:

If the need is to control only CIRCULAR cross sectional elements in a composite relationship to the datum axis, RFS, e.g., multidiameters on a shaft, use

CIRCULAR RUNOUT ↗   **EXAMPLE**

(This method controls any composite error effect of circularity and circular cross-sectional profile variations.)

If the need is to control the TOTAL cylindrical or profile surface in composite relative to the datum axis RFS, e.g., multidiameters on a shaft, bearing mounting diameters, etc., use

TOTAL RUNOUT ↗↗   **EXAMPLE**

(This method controls any composite error effect of circularity, cylindricity, straightness, angularity, and parallelism.)

NOTE   Runout is always implied as an RFS application. It cannot be applied on an MMC basis, since an MMC situation involves functional interchangeability or assemblability (probably of mating parts), in which case POSITION tolerance would be used. See below.

If the need is to control the total cylindrical or profile surface and its axis in composite location relative to the datum axis on an MMC basis, e.g., on mating parts to assure interchangeability or assemblability, use

POSITION ⊕   **EXAMPLE** | ⊕ | ∅.002 Ⓜ | A Ⓜ |

If the need is to control the *axis* of one or more features in composite relative to a *datum axis*, RFS, e.g., to control balance of a rotating part, use

CONCENTRICITY ◎   **EXAMPLE** | ◎ | ∅.002 | A–B |

NOTE   Concentricity is always implied as an RFS application. Variations in size (departure from MMC size, out-of-circularity, out-of-cylindricity, etc.) do not in themselves conclude *axis* error.

POSITION ⌖

POSITION TOLERANCING (MMC) OF COAXIAL FEATURES

This illustration shows a common application of position tolerancing of coaxial features. A functional MMC relationship is desired.

A functional datum is selected and the feature is specified to be "at true position (coaxial) within ⌀ .001 tolerance zone with feature at MMC with respect to datum A at MMC."

A functional gage (or equivalent technique) will always be used on this part. The interpretation and the illustration gage show how the part remains functionally acceptable as the concerned features depart from their MMC (worst condition) size within their size tolerance range. Further, the advantages of position tolerancing are realized since greater position tolerance is permitted as the feature and datum sizes depart from MMC. Functionally good parts are always accepted on this basis; as in all types of MMC position tolerance, size tolerance *and* position tolerance are considered together.

Positional tolerance controls axis or center plane displacement. This part illustrates the manner in which position of coaxial features incorporates the same considerations as conventional positional tolerancing of hole patterns. The axes provide the common denominator needed to relate displacement of one feature to another and to allow calculations and gage determinations based on functional design requirements. The functional gage will accommodate permissible axis displacement in terms of the surface configuration, including both form and position errors. The functional gage checks *position* only. The hole high size limit and low size limit (if necessary) are checked separately.

Zero tolerancing on position of coaxial features may occasionally be useful.

COAXIAL FEATURES

AS DRAWN

⌀1.000 +.000/−.002

-A-

⌀.500 +.000/−.002

⊕ | ⌀.001 Ⓜ | A Ⓜ

SYMBOL MEANING

⊕ | ⌀.001 Ⓜ | A Ⓜ ← WITH RESPECT TO DATUM A AT MMC SIZE

└ WITHIN ⌀.001 TOL ZONE WITH FEATURE AT MMC SIZE

└ THIS FEATURE IS TO BE LOCATED AT TRUE POSITION (COAXIAL)

MEANING

DATUM AXIS

└DATUM A

AXIS OF DATUM & FEATURE COAXIAL (.000 BASIC)

⌀.005 TOL ZONE WHEN FEATURE ⌀ AT .498 LMC, DATUM AT .998 LMC

⌀.001 TOL ZONE WHEN FEATURE ⌀ AT .500 MMC, DATUM AT 1.000 MMC

⌀.003 TOL ZONE WHEN FEATURE ⌀ AT .498 LMC, DATUM AT 1.000 MMC

GAGE

1.000   .501

*GAGE MAKERS TOL APPLY

COMPARISON BETWEEN COAXIAL FEATURES
SPECIFIED RFS (RUNOUT) AND COAXIAL FEATURES
SPECIFIED MMC (POSITION)

In a mating part situation, positional tolerancing of coaxial features can be used to good advantage. In this illustration we compare a part with coaxial features dimensioned and toleranced using total runout (RFS) with the same part dimensioned and toleranced using position (MMC).

The example at the top shows a part similar to that of the previous example except that total runout (RFS)tolerancing is used. Note that the .500 $\pm^{.000}_{.002}$ diameter in the top example has a tolerance box which states "this feature is to be within .001 total runout when mounted on datum A." This is an RFS application where the stated runout tolerance of .001 is the total allowable tolerance FIM (TIR, FIR) regardless of the size to which the .500 diameter and 1.000 datum diameter are produced. The meaning is shown at the right; the probable method of checking is also illustrated. It is seen that the dial indicator reading of the rotated .500 diameter with respect to the 1.000 datum diameter and its axis will entirely determine the runout reading and thus the acceptance or rejection of the part. The dial indicator reading must be .001 or less for part acceptance, and this tolerance is maximum.

A closer study of this part and its function may reveal that it has a *mating part* into which it must fit. Knowing this, we can determine the control of the dimensions and tolerances of the *two* parts as they are related to each other, to ensure proper fit and economic production. Where this problem of assemblability or functional requirement exists, position tolerancing and its associated advantages should be considered.

Note that the same part is illustrated under the MMC (SPECIFIED AS POSITION) subheading. The dimensions are identical. The position characteristic symbol is, however, indicated with a ϕ .001 tolerance requirement and is specified as an MMC application. At the right is the mating part containing the holes which must fit the shaft at the left. It also utilizes position control with a tolerance requirement of ϕ .002 at MMC.

Figures 1, 2, and 3 illustrate the mating part positional relationships and the manner in which additional tolerance is achieved dependent on part size variations.

In Fig. 1(a), the .001 tolerance zone permits a .0005 axis displacement of the .500 diameter on the shaft with respect to the 1.000 datum diameter A axis. On the mating part (Fig. 1b), the .002 tolerance zone on the hole permits a .001 axis displacement of the .501 hole with respect to the 1.002 datum hole A axis. This displacement includes all the errors of form and position of the feature which have an effect. Actually, in the use of MMC position, the reference to axes is primarily theoretical as a common denominator for comparison and calculation. In reality, the functional gage (or technique) which represents each mating part and its fit determines acceptance of the part on the basis of surface contact as related to the axis displacement.

The sectional view in Fig. 1(b), showing the shaft inserted into the hole, illustrates the fit of the two parts at MMC size with the .500 shaft diameter and the .501 hole diameter displaced in opposite directions to the full extent of the position tolerance. The parts assemble satisfactorily.

RFS (SPECIFIED AS TOTAL RUNOUT)

Ø1.000 +.000 −.002

−A−

Ø.500 +.000 −.002

↗↗ .001 A

MEANING

DATUM DIA A

DATUM AXIS

.001 FIM (TIR, FIR)

VEE BLOCK

.001 WIDTH TOL ZONE FIM (TIR, FIR)

MMC (SPECIFIED AS POSITION)

Ø1.000 +.000 −.002

−A−

Ø.500 +.000 −.002

⏀ Ø.001 Ⓜ A Ⓜ

MATING PART

Ø1.002 +.002 +.002 −.000

−A−

Ø.501 +.002 −.000

⏀ Ø.002 Ⓜ A Ⓜ

POSITION TOLERANCE ZONES AT MMC

1.000 MMC DATUM

.500 MMC

.0005 DISPLACEMENT OF .500 DIA AXIS

.001 DIA TOTAL TOL ZONE MMC

1.002 MMC DATUM

.002 DIA TOTAL TOL ZONE MMC

.501 MMC

.001 DISPLACEMENT OF .501 DIA AXIS

FIGURE 1   **(a)**   **(b)**

POSITION TOLERANCE ZONES AT LEAST MATERIAL CONDITION

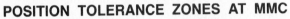

.998 DATUM

.498

.0025 DISPLACEMENT OF .498 DIA AXIS

.005 DIA TOTAL TOL ZONE

1.004 DATUM

.006 DIA TOTAL TOL ZONE

.003 DISPLACEMENT OF .503 DIA AXIS

.503

FIGURE 2   **(a)**

TOL ZONE & SHAFT DIAS

TOL ZONE & HOLE DIAS

SHAFT AT MMC
HOLES AT LST. MATL CONDITION

TOL ZONE & SHAFT DIAS

TOL ZONE & HOLE DIAS.

SHAFT & HOLES AT LST. MATL CONDITION

TOL ZONE & SHAFT DIAS

TOL ZONE & HOLE DIAS

HOLES AT MMC
SHAFT AT LST. MATL CONDITION

FIGURE 3   **(a)**   **(b)**   **(c)**

In this situation the part shaft and hole sizes are at MMC; that is, we have the largest permissible shaft size and the smallest permissible hole size. If the parts assemble under these conditions, it is reasonable to assume that as the shaft gets smaller and the hole larger, the tolerance could be increased while continuing to permit assembly.

Figure 2(a) illustrates how the position tolerance zone increases as the part and hole sizes depart from MMC to the opposite size extreme (LEAST MATERIAL CONDITION). Note that the .500 shaft diameter is now .498 and the 1.000 datum is now .998, or the low limit of the shaft diameter sizes. In Fig. 2(b) the .501 hole diameter is now .503 and the 1.002 datum diameter is now 1.004, or the high limit of the hole sizes.

It may now be seen that the reduction in shaft sizes and increase in hole sizes permit a greater displacement of the mating features, resulting in a permissible ϕ .005 total position tolerance zone on the stepshaft at the left and a ϕ .006 total position tolerance zone on the mating part at the right.

The sectional view at the far right showing the shaft inserted into the hole illustrates the fit. The parts still assemble satisfactorily.

With the use of position, the permissible tolerance has been increased from ϕ .001 to ϕ .005 on the shaft at the left and from ϕ .002 to ϕ .006 on the mating part at the right.

Referring to the part examples under the MMC (POSITION) subheading, note that a rule-of-thumb method of determining the maximum possible position tolerance is to *add* the stated tolerance found in the symbol box and the total size tolerance of the features involved. For example, on the shaft at left, the position tolerance of ϕ .001 found in the symbol box, plus the .002 total size tolerance of the .500 shaft diameter, plus the .002 total size tolerance of the shaft 1.000 datum diameter (or .001, plus .002, plus .002) equals .005 total. On the mating part at the right, the position tolerance of ϕ .002 found in the symbol box, plus the .002 total size tolerance of the .500 hole, plus the .002 total size tolerance of the 1.002 datum hole diameter (or .002, plus .002, plus .002) equals .006 total.

Remember, however, that the resultant tolerance is dependent on the actual size to which the features are produced. Thus the tolerance under an MMC application is usually somewhere between the stated tolerance and the opposite limit as determined by actual feature sizes.

Figure 3 illustrates the assembly relationship of the shaft and holes at various extremes. Figure 3(a) shows the shaft at MMC or its largest size, with the mating part holes at LEAST MATERIAL CONDITION, or their *largest* size. Figure 3(b) shows both parts at LEAST MATERIAL CONDITION or at the size limits opposite to the calculated MMC sizes; that is, the shaft is at its smallest allowable size and the holes are at their largest allowable size. This is the condition illustrated in Fig. 2; it allows the greatest amount of tolerance on both parts. Figure 3(c) shows the holes at MMC or their smallest size and the shaft at LEAST MATERIAL CONDITION or its smallest size. Note that the parts assemble in each case.

Numerous other combinations are, of course, also possible and acceptable, so long as the established size and location tolerances are held.

These examples again illustrate that in position tolerancing, size tolerances affect position tolerances. It is advantageous to consider this fact whenever possible so that maximum tolerance yield may be achieved.

RFS (SPECIFIED AS TOTAL RUNOUT)

MEANING

MMC (SPECIFIED AS POSITION)

MATING PART

POSITION TOLERANCE ZONES AT MMC

FIGURE 1   (a)   (b)

POSITION TOLERANCE ZONES AT LEAST MATERIAL CONDITION

FIGURE 2   (a)   (b)

FIGURE 3   (a)   (b)   (c)

MMC CALCULATIONS TO DETERMINE TOLERANCE

The actual calculations necessary to determine position tolerances of coaxial features of mating parts are illustrated in this example. The mating parts are the same as those in the preceding example.

Note that we have selected functional datums: the diameters of mating features, the 1.000 shaft diameter, and the 1.002 hole diameter. This establishes a common datum axis and the relationship between the diameters of the two parts. This information is necessary for our calculation.

Selected *clearances* of the holes to the mating shaft diameters are established by the design requirements. These clearances in turn establish the position tolerances to be stated for these features.

Note that the first step illustrated under the subheading POSITION TOLERANCE CALCULATIONS shows the .500 MMC shaft size of part 2 subtracted from the .501 MMC hole size of part 1, resulting in a difference of .001. Also, the 1.000 MMC datum shaft size of part 2 is subtracted from the 1.002 MMC datum hole size of part 1, resulting in a difference of .002.

The .001 result of the first calculation and the .002 result of the second calculation are added to give the .003 total combined position tolerance for both parts and for their interrelated features. This total tolerance is then divided to establish the required position tolerance on each individual part. The division can be made in any combination, so long as the individual part tolerances total the calculated combined tolerance, in this case, .003.

For this example, .002 has been selected as the position tolerance for the .501 hole in part 1, and .001 as the position tolerance for the .500 diameter shaft of part 2. These two figures total .003 and comply with the allowable total combined position tolerance.

Dependent on feature size variation, we may realize an extra or bonus tolerance.

This possible additional position tolerance as the feature sizes depart from MMC size is calculated as shown on the lower portion of the illustration.

Remember, however, that this increase in position tolerance is dependent upon the *actual* sizes to which the features are produced within their size tolerances.

Since most parts are produced somewhere between the high and low size tolerance extremes, the actual position tolerance permissible on part number 1 would be somewhere between .002 and .006, say approximately .004. It could, however, be .006 and still provide the proper fit to the mating part. The actual position tolerance permissible on part 2 would probably be somewhere between .001 and .005, say approximately .003. It could, however, be .005 and still provide the proper fit to the mating part.

The advantages of using position tolerances on coaxial features are now apparent.

1.   Greater tolerance is permitted.
2.   The fit of the mating part is guaranteed when the dimension and the tolerance requirement are calculated simultaneously and consideration is given to the possible variables which can occur in the assembly relationship.

The method of calculating position tolerances described above assumes the possibility of a zero interference—zero clearance condition of the mating part features at extreme tolerance limits. Additional compensation of the calculated tolerance values should be considered as necessary relative to a particular application.

POSITION TOLERANCE—COAXIAL FEATURES—MATING PARTS

AS DRAWN

PART 1

MATING PARTS

$$T = \frac{(H - S) + (D_1 - D_2)}{2}$$

H = HOLE PART 1
S = SHAFT PART 2

D_1 = DATUM PART 1
D_2 = DATUM PART 2

$2T = T_1 + T_2$

PART 2

POSITION TOLERANCE CALCULATIONS

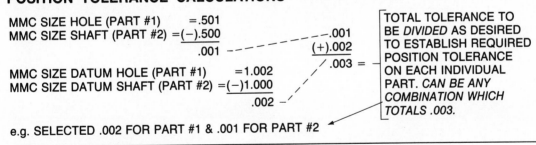

MMC SIZE HOLE (PART #1)   = .501
MMC SIZE SHAFT (PART #2) = (−).500
                                              .001 ------ .001
                                                      (+).002
                                                  .003 =
MMC SIZE DATUM HOLE (PART #1)   = 1.002
MMC SIZE DATUM SHAFT (PART #2) = (−)1.000
                                                   .002

⌐ TOTAL TOLERANCE TO
  BE *DIVIDED* AS DESIRED
  TO ESTABLISH REQUIRED
  POSITION TOLERANCE
  ON EACH INDIVIDUAL
  PART. *CAN BE ANY
  COMBINATION WHICH
  TOTALS .003.*

e.g. SELECTED .002 FOR PART #1 & .001 FOR PART #2

EXTRA TOLERANCE FOR EACH PART

PERMISSIBLE HOLE POSITION TOL AS HOLE SIZES DEPART (GET LARGER)
FROM MMC:

STATED POSN TOL WITH HOLE AT .501 MMC	=.002
PLUS TOTAL .501 HOLE TOL	+.002
POSN TOL WITH DATUM HOLE 1.002 AT MMC	.004
PLUS TOTAL 1.002 DATUM HOLE TOL	+.002
PART 1 TOTAL POSN TOL WITH BOTH HOLES AT LEAST MAT'L CONDITION	=.006

PERMISSIBLE SHAFT POSITION TOL AS SHAFT SIZES DEPART (GET
SMALLER) FROM MMC:

STATED POSN TOL WITH SHAFT AT .500 MMC	=.001
PLUS TOTAL .500 DIA TOL	+.002
POSN TOL WITH DATUM SHAFT DIA 1.000 AT MMC	=.003
PLUS TOTAL 1.000 DATUM DIA TOL	+.002
PART 2 TOTAL POSN TOL WITH BOTH SHAFT DIAS AT LEAST MAT'L CONDITION	=.005

MMC GAGES

Positional tolerances may, of course, be used on any part, even though there is no direct assembly relationship between mating part features. The tolerance advantages and the possibility of using functional or receiver gages for effective and economic inspection often makes the position technique very desirable under such conditions. However, where the assemblability of mating part features is involved, position tolerancing is always recommended. Position tolerancing of coaxial features of mating parts is an ideal example of this use.

Functional or receiver gages for coaxial features of mating parts simulate the fit of the actual part features in a manner similar to the hole pattern functional gages. As in the hole pattern functional gage, part feature size departure from MMC permits greater part acceptance based on the functional interrelationship of size and position variations.

The illustration shows the two mating parts previously discussed and examples of functional gages to check the position requirements of each part. The upper illustration, part 1, is shown with a gage for functionally checking the position tolerance of the holes. The simple calculations to determine the gage dimensions are shown at the right. Gage-maker's tolerances also apply as required.

The lower illustration shows part 2 and a functional gage for checking the position of the shaft diameters of this part. The calculations are shown at right. Gage-maker's tolerances also apply as required.

The illustrated gages will check only the position tolerances of the part. Sizes of the associated features require a separate size check.

PART 1

Ø1.002
+.002
−.000
−A−

Ø.501 +.002
−.000

⊕ | Ø.002 Ⓜ | A Ⓜ

POSITION FUNCTIONAL GAGE

*GAGE MAKERS TOL AS REQUIRED

Ø1.002*

Ø.499*

GAGE CALCULATIONS
MMC SIZE PART HOLE   =.501
MINUS STATED
    POSITION TOL   −.002
                GAGE DIA =.499

MMC SIZE OF PART
DATUM HOLE DIA. (1.002)
ESTABLISHES LARGE
GAGE DIA.   1.002

PART 2

Ø1.000 +.000
−.002
−A−

Ø.500 +.000
−.002

⊕ | Ø.001 Ⓜ | A Ⓜ

POSITION FUNCTIONAL GAGE

Ø1.000*   *Ø.501

*GAGE MAKERS TOL AS REQUIRED

GAGE CALCULATIONS
MMC SIZE PART SHAFT
                DIA   .500
PLUS STATED
    POSITION TOL   +.001
    GAGE HOLE DIA =.501

MMC SIZE OF PART
DATUM SHAFT DIA (1.000)
ESTABLISHES LARGE
GAGE HOLE DIA   1.000

RELATION TO DATUM SURFACES USING A RADIAL HOLE PATTERN

Radial hole patterns may be controlled by position tolerancing in a manner similar to that used on more conventional hole patterns.

Since the four hole pattern in the example at the right requires location as well as feature-to-feature relationship (in-line), composite positional tolerancing may be used to advantage.

Datums are established relative to the part function, the face surface A and the .900 diameter, datum B. Imagine this part mating with another at assembly on the surface A and located and centered on B so pins will pass into the holes. MMC is desired so the hole and mating part pin sizes are used to determine the hole-to-hole position tolerances.

In this example, the holes are desired to be within one plane (coplanar), in-line (colinear), and spaced 90° apart. Therefore, as a pattern they may vary from that basic configuration only to the extent of permissible tolerances. In this case, the pattern as an entity (established from the holes themselves) may vary from position (and parallel to A) within .030 diameter at MMC. In the pattern, the holes may vary from their basic colinear alignment within .005 diameter at MMC. The axes of the four holes must lie within both the pattern locating tolerance zones and the feature relating tolerance zones.

In terms of gaging this part, two separate gages or gaging operations would be required: one to orient to datums A and B and evaluate the .030 diameter tolerance, and another to evaluate the .005 diameter tolerance. The latter could be simply accomplished by passing a gage pin of .120 diameter through two holes in-line simultaneously. The hole size requirements must, of course, also be met.

As the four holes depart from MMC, more position tolerance is acquired (as has been previously described in other examples). In this case, although the example is more complex, the principles are the same.

AS DRAWN

MEANING

PERMISSIBLE HOLE
AXIS VARIATION

DATUM
AXIS B

90°

SECONDARY DATUM
B (.900 MMC)

PARALLEL

.750

PRIMARY DATUM PLANE A

⌀.005 FEATURE RELATING TOL ZONE (AT MMC)
⌀.008 (AT LMC)
⌀.03 PATTERN LOCATING TOL ZONE (AT MMC)
⌀.033 (AT LMC)

THE AXIS OF THE 4 HOLES MUST LIE WITHIN
BOTH THE PATTERN LOCATING TOL ZONES
(⌀.03 AT MMC) AND THE FEATURE RELATING
TOL ZONES (⌀.005 AT MMC)

POSITION
EXTENDED
PRINCIPLES

SEPARATE PATTERNS OF HOLES

Patterns of holes (or features) on a part which are either shown by common basic or untoleranced dimensions or are related to common datums are normally assumed to be related and therefore considered a single composite pattern. Occasionally, convenience in the placement of dimensions or the need to show common dimension origination from related datum features on the drawing can unintentionally impose restrictions on production or inspection.

If certain holes (or features) in a BASIC dimensioned pattern are related to one another functionally but not to others which are also shown in the same pattern, or if we wish to produce or gage certain portions of the pattern separately, we may indicate this by the notation SEP REQT or SEPARATE REQUIREMENT beneath or adjacent to the concerned feature control frame or frames. Coding the appropriate holes ⌖ on the drawing is helpful in clarifying the drawing intent.

Hole (or feature) patterns on a part separated by rectangular coordinates, related to *different* datums, or having no direct tie by dimensional arrangement, are considered to be separate patterns; no further notation is required in this case.

PERPENDICULARITY (SQUARENESS) CONTROL WITHIN POSITION TOLERANCING

Where out-of-perpendicularity of threaded or plain holes could cause inserted screws, bolts, studs, pins, or dowels to interfere with mating parts or result in cocked seating of screw or bolt heads, closer control of perpendicularity *within* the position tolerance may be required.

Perpendicularity control may be specified on the drawing, using one of the methods illustrated below.

PERPENDICULARITY SPECIFICATION ON DRAWING

AS DRAWN   **MEANING**

CALCULATION TO COMPENSATE POSITIONAL TOLERANCE FOR OUT-OF-PERPENDICULARITY OR WHERE PROJECTED TOLERANCE ZONE NOT USED:
WHERE: D = MINIMUM DEPTH OF HOLE OR PART THICKNESS FOR FIXED FASTENER (OR PIN)
          P = MAXIMUM THICKNESS OF PART WITH CLEARANCE HOLE OR PROJECTION OF FASTENER (OR PIN)
TO CALCULATE POSITION TOLERANCE:

$$T = \frac{H-F}{\left(1 + \frac{2P}{D}\right)}$$

TO CALCULATE CLEARANCE HOLE MMC SIZE:

$$H = F + T\left(1 + \frac{2P}{D}\right)$$

NOTE   Where the projected error of inserted screws or pins in such holes could cause mating part interference, the holes in the mating part may have to be enlarged to account for this tolerance variation unless the "projected tolerance zone" method is applied to the tapped holes.

PROJECTED TOLERANCE ZONE

The projected tolerance zone method prevents the condition shown in Fig. 1, where interference could possibly exist with conventional positional tolerancing. The variation from perpendicularity of the portion of the bolt passing through the mating part is of concern. Therefore the location and perpendicularity of the tapped hole is of importance insofar as it affects this extended portion of the bolt. The projected tolerance zone method (Fig. 2) eliminates this interference.

With this method, we can use conventional "fixed fastener" calculations to determine the position tolerance. Further, specifying by this method means that gaging techniques will simulate the mating part relationship, and the projected perpendicularity error will be accounted for in the tolerance and in the gaging.

FIGURE 1

FIGURE 2

EXAMPLE   **MEANING**

ROUND HOLES–0 TOLERANCE (PERFECT POSITION AT MMC)

"Zero" (0) position tolerancing is a technique adaptable to situations requiring functional interchangeability and maximum tolerance advantage in the feature size, form, orientation, and position interrelationships. Where mating parts and features are simply to mate up or "GO" and tangent contact of the mating features could occur, zero tolerancing is technically acceptable.

However, under some conditions, zero position tolerancing is *not* appropriate. For example, where specific running clearances, fit, or similar special mating feature conditions are required, zero position tolerancing will not, in general, be technically applicable. There are other considerations, also, which require evaluation to determine whether or not zero position tolerancing is applicable.

For most position applications and examples shown elsewhere in this text, we could have used the zero position tolerancing method. It is an optional method for stating many common position mating part requirements.

As foregoing sections of this text have emphasized, position tolerances are usually established on the basis of MMC size relationships of mating part features. The feature sizes are the criteria with which the process of developing the position tolerances starts. The designed clearance between the mating components is the basis for the position tolerances which are stated on the drawing and applied in manufacture. When the features specified by position tolerances are actually produced, any size departure from the MMC size (e.g., enlarging size of a hole) adds to the permissible position tolerance.

In zero position tolerancing the same principles apply, except that the position tolerance stated is always a fixed "zero," with all the tolerance placed on the size dimension. This, of course, assumes that the actually produced feature will show some deviation from MMC, which is then added to the "zero" tolerance to give a working position/orientation tolerance.

In either conventional or zero position tolerancing, size, form, orientation, and position variations are considered simultaneously as a composite value. This is really the fundamental principle (along with the MMC principle) on which functional position tolerancing is based. The reason for this is that related mating part features perform their function in the space limitations provided, regardless of whether that space is derived from size, form, orientation, or position variation.

There is some controversy about zero position tolerancing, with both proponents and opponents contributing valuable comments. However, a good understanding of the principles involved, their proper application, and an awareness of their limitations will go a long way toward ensuring proper use of this effective geometric tolerancing tool.

A forceful argument for the use of zero tolerancing arises in situations where a produced part with a positional hole pattern might be acceptable to a functional gage yet be rejectable on the basis of a low limit "GO" size violation, with the result that functionally good parts might be scrapped.

At this point, we wish to emphasize that in conventional position tolerancing, the stated size tolerance can be used for size, form, orientation, and position variations as the feature size departs from MMC, whereas a stated position tolerance may be used only for orientation and position variations. Size tolerance variations of the feature from MMC size can thus add to the position tolerance; but, according to standard practices, unused position variations cannot be added to size tolerances.

The above principle is best described by referring to the CONVENTIONAL POSITION TOLERANCE APPLICATION example which follows in this section of the text. The notation at the bottom of the illustration states that if the hole is produced in perfect location, its size will be permitted to exceed the low limit .255 (MMC) size down to the virtual condition size of .250. The virtual condition size is developed from the MMC size of the hole, .255, minus the stated position tolerance, .005. This is, of course, also the functional gage pin size, and represents the mating part feature at its extreme condition of assembly.

Obviously, not many parts will be produced in perfect position. However, the point is that when the position tolerance, or any portion of it, has *not* been used up, the *size* of the feature should reap the benefit (e.g., the hole should be allowed to get smaller an equal amount and yet be accepted as functional).

Note that the virtual condition size developed in the ZERO POSITION TOLERANCE example is also .250. In this method the virtual condition size and the MMC size are the *same*. The .005 position tolerance of the ''conventional'' example has been shifted to the size tolerance in the ''0'' method. The result for both parts, insofar as resulting position gaging is concerned, is the same; but the above-mentioned ''rejectable'' part (low limit size violation) will be acceptable under the zero method.

The .250 diameter, or virtual condition size, on both examples represents the most extreme condition of assembly at MMC, or the most extreme mating condition of the hole which would accept the mating part feature. Thus, the zero method recognizes, accounts for, and uses the full tolerance advantage. Size tolerances, in particular, are fully utilized with no loss of form, orientation, or position tolerance.

Further analysis of zero tolerancing, however, provokes questions that tend to temper some of its advantages. First, for the less experienced or uninitiated user, zero tolerancing presents a psychological barrier: the zero may give a false impression of the ''perfection'' expected. Second, the designer may feel that he is relinquishing excessively broad discretion to the production departments, thus abdicating design responsibilities in favor of production conveniences such as larger size tolerances. Other comments heard from those hesitant to make full use of zero tolerancing are: it will be misunderstood, it costs more, it is only for gaging convenience, it may result in line-to-line fits.

These objections are primarily due to lack of understanding and experience. Improved understanding and repeated experience with practical applications will readily convince the user of the advantages of zero tolerancing.

Zero form tolerancing, occasionally used on perpendicularity, is not discussed here. However, examples of it are given in appropriate sections of the text. The reasoning underlying zero *orientation* tolerancing is the same as that on which zero position tolerancing is based except for the added consideration of position relationships. A position tolerance, of course, contains orientation tolerance. An understanding of the zero orientation control method, however, might contribute to better acceptance of the zero position principles.

ZERO POSITION

To clarify the reasons underlying the use of the ''zero'' position method versus the ''conventional'' position method, further discussion is presented in Figs. 1 and 2.

In Fig. 1, one of the holes illustrated in the 0 method example is shown with reference to the gage pin (or simulated mating part component). It is seen that the zero (0) position specification requires a perfect part (perfect form, orientation, and perfect position) when at MMC, or virtual condition size.

FIGURE 1

FIGURE 2

Since there must be *some* clearance between the hole and the inserted mating component or they will not assemble, there is an immediate deviation from the perfect "zero-clearance-zero interference" situation and some tolerance is acquired.

Further, since additional tolerance latitude is usually required to assure assembly, we may find it useful to establish a fixed position tolerance. Thus we calculate an acceptable tolerance on the basis of MMC sizes and use the "conventional" position method.

Figure 2 illustrates the "conventional" method and the established position tolerance. The tolerance of .005 will permit either position or orientation error (or a combination of both) to this extent, when the feature is at MMC. With the same size gage pin as in Fig. 1, we see that the position tolerance of .005, plus the size tolerance of .005, is equivalent to the .010 size tolerance obtained by the zero method in Fig. 1.

An application which will demonstrate the effectiveness of zero tolerancing is a very critical relationship of locating dowels relative to locating holes on a mating part.

As an example, imagine $.2500 \, {}^{+.0000}_{-.0015}$ on the locating dowels and $.2520 \, {}^{+.0015}_{-.0000}$ on the locating holes. Using the conventional position "fixed fasteners" method, the calculations are:

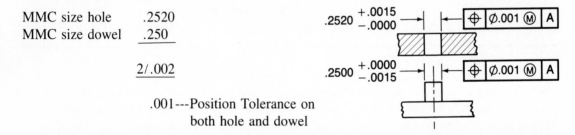

MMC size hole	.2520
MMC size dowel	.250
	2/.002

.001---Position Tolerance on
both hole and dowel

The actual position tolerance in production on both parts would be somewhere between .001 and .0025 (increase due to MMC departure). A functional gage pin size to check the holes would be .251 (hole MMC .252, MINUS POSN TOL .001 = .251). Since the gage pin represents the worst condition (virtual condition size) of the mating dowel at .251, the hole size could be acceptable functionally at .251; yet this exceeds the stated hole size low limit.

The dowel size, too, could be functional at .251 (dowel MMC .250, PLUS .001 = .251) which represents the mating part hole at the extreme condition (virtual condition size). This exceeds the stated dowel size high limit.

However, the zero position method can provide more total tolerance and yet guarantee proper control if stated as:

$$.2510 \, {}^{+.0000}_{-.0025} \text{ (dowel)} \quad \text{and} \quad .2510 \, {}^{+.0025}_{-.0000} \text{ (hole)}.$$

A comparison of the two methods in terms of the full tolerance range difference between the hole and dowel which determines usable size, orientation, and position tolerance is shown on the following page.

FEATURE SIZE RANGE	USABLE SIZE TOL	USABLE POSN TOL

Conventional Method

⌀ Hole .252 to .2535   .005   (Hole) up to .0025 DIA
⌀ Dowel .250 to .2485   − .002   (Dowel) up to .0025 DIA
        .002 to .005   ⌀ .003 USABLE

↙ LOSE .002 SIZE TOL: ↗

Zero Method

⌀ Hole .251 to .2535   ⌀ .005 USABLE   (Hole) up to .0025 DIA
⌀ Dowel .251 to .2485   (Dowel) up to .0025 DIA
        .000 to .005

↙ FULL UTILIZATION OF SIZE TOL ↗   SAME POSN TOL RESULTS

The foregoing example assumes that there is only a remote probability that both hole and dowel will actually be produced at exactly .251; hence some clearance will be present. However, if this should cause concern and some compensation is desired, a slight adjustment of specified size limits (and virtual condition size) can be made to eliminate the problem and ensure that no line-to-line condition will result.

As an alternative to the zero position tolerancing method, some companies have established in-house standards which permit the use of a functional gage as a hole "GO" size gage, thus allowing, for example, the low limit size of a hole to deviate simulating the zero method and the mating part relationship. However, transfer of work to outside sources would then require special documentation to assure that specifications are properly interpreted.

The illustrations on the following pages further illustrate the principles described on page 192. Note that the "conventional" method establishes a fixed position tolerance which presents the possibility of a part being *acceptable* to a functional gage but *rejectable* due to exceeding the low limit (MMC) size. The zero method on page 196, however, eliminates that problem. It does, however, retain the inherent considerations previously discussed. Use of zero position tolerancing should be based upon the best balance of design/production objectives.

CONVENTIONAL POSITION TOLERANCE APPLICATION FOR COMPARISON WITH ZERO POSITION TOLERANCE

POSITION TOLERANCE (CONVENTIONAL)

AS DRAWN

4X Ø.255 +.005 −.000

⊕ | Ø.005 Ⓜ | A | B | C

SYMBOL MEANING

⊕ | Ø.005 Ⓜ | A | B | C ← WITH RE-SPECT TO DATUMS A, B, C

— WITHIN Ø.005 TOL ZONE AT MMC

— THESE FEATURES TO BE AT TRUE POSITION

ACTUAL FEATURE Ø SIZE	Ø TOL
.255 (MMC)	.005
.256	.006
.257	.007
.258	.008
.259	.009
.260 (LMC)	.010

GAGE PIN SIZE = .250 (MMC .255 MINUS .005 TOL)

MEANING

.005 TOTAL UNUSED POSITION TOL

Ø.255 MMC HOLE (LOW LIMIT)

HOLE POSITION PERFECT

PART

Ø.250 GAGE PIN (VIRTUAL CONDITION SIZE)

ASSUMING THAT THE GAGE PIN REPRESENTS THE MOST EXTREME MATING CONDITION, AS POSITION APPROACHES PERFECT, IT IS EVIDENT THAT THE HOLE SIZE COULD GO DOWN TO .250 (.005 BELOW .255 LOW LIMIT OF HOLE) AND STILL PASS THE GAGE PINS, HOWEVER, PARTS BELOW THE LOW (MMC) LIMIT HOLE SIZE OF .255 WOULD BE REJECTED ON SIZE YET ARE GOOD PARTS.

ACTUAL FEATURE ∅ SIZE	POSN. ∅ TOL
.250 (MMC)	.000
.251	.001
.252	.002
.253	.003
.254	.004
.255	.005
.256	.006
.257	.007
.258	.008
.259	.009
.260 (LMC)	.010

ADVANTAGES OF ZERO POSITIONAL TOLERANCE

1. *Hole "GO" plug gage not needed* if using "GO" functional gage which checks positional location of pattern and hole low limit size simultaneously.

2. *No unused positional tolerance* when using zero positional tolerance method. As locations approach perfection under a conventional (specified pos tol e.g., .005) position application, the unused positional tolerance *cannot* be added to the *size* tolerance. Therefore, under some conditions good parts are rejected as they exceed the low limit size.

COAXIAL FEATURES–ZERO TOLERANCE

Zero position tolerancing as illustrated in the preceding examples may also be applied to coaxial features.

Where position relationships between coaxial features must be very exact, zero tolerancing may provide the necessary control. As in other position applications, it is appropriate for mating parts. It has the advantage of fully utilizing *size* tolerances as these interrelate with form, orientation, and position relationships.

The example on the next page shows zero tolerancing applied to a simple part. Position tolerance is acquired as the feature sizes (feature and datum feature) depart from MMC. When the part is everywhere at its MMC size, it must be perfectly located. However, in normal applications, the sizes produced are somewhere within the size tolerance range and thus develop allowable positional tolerance. The actually produced sizes establish the position tolerance to the amount of size departure from MMC.

Note that the sizes of both the feature *and* the datum have an effect on the total relationship and the developed tolerance. Just as in all applications of position tolerancing, the part must be produced and its size established before the position tolerance is fixed.

A functional or receiver gage is also illustrated. The principles involved are identical to those in the previous examples. However, with the zero method, the high limit "GO" size of the diameter is also checked, along with the position requirement.

For this application, the use of maximum size tolerances and functional principles has been facilitated by the zero method.

AS DRAWN

Ø1.000 +.000 / −.002

-A-

Ø.500 +.000 / −.002

⊕ | Ø.000 Ⓜ | A Ⓜ

SYMBOL MEANING

⊕ | Ø.000 Ⓜ | A Ⓜ ← WITH RESPECT TO DATUM A AT MMC

WITHIN Ø.000 TOL ZONE AT MMC

THIS FEATURE TO BE AT TRUE POSITION

MEANING

DATUM AXIS

DATUM A

AXES OF DATUM & FEATURE COAXIAL (.000 BASIC)

Ø.004 TOL ZONE WHEN FEATURE AT LOW (LMC) LIMIT .498, DATUM AT LOW (LMC) LIMIT .998

Ø.000 TOL ZONE WHEN FEATURE Ø AT .500 MMC, DATUM AT 1.000 MMC

Ø.002 TOL ZONE WHEN FEATURE Ø AT LMC LIMIT .498, DATUM AT 1.000 MMC

GAGE

Ø1.000*   Ø.500*

*GAGE MAKERS TOL APPLY

MATING PARTS, FIXED FASTENERS

The following series of figures further illustrates mating part situations and the use of the fixed fastener method of calculation. It covers the typical applications listed below.

MATING PARTS, FIXED FASTENERS

CALCULATIONS
(BASE TO COVER)

.138–32 SCREW
    MMC = .138
COVER HOLE
    MMC = .148
    2|.010
    ∅.005
ON EACH

BASE BOSS
    MMC = .151
COVER
HOLE MMC = .153
    2|.002
ON EACH ∅.001

BASE

CALCULATIONS
(BASE TO GASKET)

.138–32 SCREW
    MMC = .138
GASKET
HOLE   = .154
    MMC .016

.005 ALLOWED ON
.138–32 HOLE
BASE
(.016–.005 = .011)

.011 ALLOWED
ON GASKET

GASKET

BASE BOSS
    MMC = .151
GASKET
HOLE MMC = .164
    .013

.001 ALLOWED
  FOR BASE
  BOSS
(.013 .001 = .012)
.012 ALLOWED
ON GASKET

COVER

BASE

NOTE: ALL FEATURE RELATING
TOLERANCES ARE SINGLE PATTERN

COVER

CALCULATIONS	AIR VENT HOLES TO	TO CLEAR = .580 DIA.
MMC .138−32 .138	LINE UP TO MINIMUM	MMC HOLE = .594 DIA.
MMC HOLE .148	DIA. OF .146	.580 MIN. DIA.
2⌐.010	MMC HOLES .157	Ø.014
Ø.005	MIN DIA. .146	
	Ø.011	ON EACH PART
.005 ON EACH PART		
(COULD BE .006 & .004,	ON EACH PART	
.007 & .003, ETC.)		

Imagine .146 & .580 dias as columns of air (floating fasteners). Floating fastener
method may be used since overlap of edges is permissible; one dia need not clear the
other so long as the .146 & .580 dias will pass.

MATING PARTS AND FIXED FASTENERS

CALCULATIONS
MMC .164−32 .164
MMC HOLE   .174
―――――――
Ø.010
USE .010 TP TOL
ON .164−32 HOLES

USE .000 ON
COVER HOLES
(COULD USE .005 &
.005, .006 & .004,
ETC.)

HOUSING

COVER

-C-

1.100

.980   .980

-B-

.900

.400

1.800 ±.010

.200

2.200 ±.010   .150 ±.010   1.200 ±.010

-A-

CALCULATIONS
MMC .164 – 32   .164
MMC HOLE   .172
        ─────
        ⌀.008

USE .008 TP ON
.164 – 32 HOLE

.000 ON .172 HOLE

(COULD USE .004
& .004, .006 &
.002, ETC.)

MMC BOSS   .130
MMC HOLE   .130
        ─────
TP ON BOTH .000

TP TOL FOR
MATING BOSSES
& HOLES DE-
RIVED FROM
SIZE DEPARTURE
FROM MMC

2x ⌀.130 $^{+.000}_{-.005}$

| ⊕ | ⌀.000 Ⓜ | A | B | C |

.164 – 32 UNC-2B
▽ .38 ±.02

| ⊕ | ⌀.008 Ⓜ | A | B | C |
| .255 Ⓟ |

HOUSING

-C-

1.100

.980   .980

1.800 ±.010

.400

.200

.900

-B-

2.200 ±.010

-A-

.250 ±.005

2x ⌀.130 $^{+.005}_{-.000}$

| ⊕ | ⌀.000 Ⓜ | A | B | C |

⌀.172 $^{+.005}_{-.000}$

| ⊕ | ⌀.000 Ⓜ | A | B | C |

COVER

COMBINATION AND UNIQUE APPLICATIONS

Position tolerancing may be applied in numerous ways. The principles are extended in the following illustrations to unique combinations of DIA and TOTAL tolerances, square holes, etc. Note that the established principles can be adapted to many different situations.

ROUND HOLES, GREATER POSITIONAL TOLERANCE IN ONE WAY THAN IN THE OTHER—BIDIRECTIONAL

AS DRAWN

MEANING

GAGE

NOTE: Gage makers tol apply

DIAMETER AND TOTAL WIDE TOLERANCE ZONES, ROUND HOLES AND ELONGATED HOLES—BIDIRECTIONAL

AS DRAWN

2X .300 ±.005

| ⊕ | .025 Ⓜ | A | B | C |

2X .250 ±.003

| ⊕ | .005 Ⓜ | A | B | C |

-C-

2.75 ±.02

1.75 ±.02

R

.425

.400

-B-

.375

2.000

2X ∅.250 ±.003

| ⊕ | ∅.005 Ⓜ | A | B | C |

.500 ±.003

-A-

MEANING

DATUM C   .375   2.000

DATUM A→   ½ WIDTH

.025 TOTAL WIDE ZONE AT MMC .295 HOLE LENGTH, .035 TOTAL WIDE ZONE AT LMC .305 HOLE LENGTH.

.425

90°

½ WIDTH   .005 TOTAL WIDE ZONE AT MMC .247 HOLE WIDTH .011 TOTAL WIDE ZONE AT LMC .253 HOLE WIDTH.

.400

90°

90°

DATUM B

∅.005 TOL ZONE AT MMC .247 HOLE, ∅.011 TOL ZONE AT LMC .253 HOLE

NOTE Remaining part profile tolerance zones are determined by the overall size limits

GAGE (FOR POSITION PATTERN)

.375

2.000

(FOR POSN PATTERN)
∅.242 (MMC HOLE .247 MINUS POS TOL ∅.005) (2)

SIM
DATUM C

∅.270 (MMC HOLE LENGTH .295 MINUS POS TOL .025 TOTAL WIDE ZONE) (2)

.425

∅.242 (MMC HOLE WIDTH .247 MINUS POS TOL .005 TOTAL WIDE ZONE) (2)

.400

SIM
DATUM B

SIM
DATUM A

NOTE   Gage makers tol apply

ELONGATED HOLES, TOTAL WIDE TOLERANCE ZONES— BIDIRECTIONAL

AS DRAWN

MEANING

NOTE Remaining part profile tolerance zones are determined by the overall size limits

GAGE (FOR POSN. PATTERN)

NOTE   Gage makers tol apply

PATTERN AND FEATURE ORIENTATION FROM DATUM SURFACES

USING PART PROFILE AS DATUM

AS DRAWN

MEANING

*Centers may shift within a zone equal to the departure of datum features A & B sizes from MMC.

GAGE

NOTE   Gage makers tol apply

TOTAL WIDE TOLERANCE ZONES, FEATURES TO BE SYMMETRICALLY LOCATED—BIDIRECTIONAL

AS DRAWN

MEANING

GAGE

NOTE Gage makers tol apply

SQUARE HOLES, TOTAL WIDE ZONE POSITIONAL TOLERANCE

AS DRAWN

MEANING

.005 TOTAL WIDE ZONE AT MMC .307 SQUARE HOLE, .011 TOTAL WIDE ZONE AT LMC .313 SQUARE HOLE

½ WIDTH

NOTE Remaining part profile tolerance zones are determined by the overall size limits.

GAGE (FOR POSN. PATTERN)

NOTE Gage makers tol apply

SQUARE HOLES, GREATER POSITIONAL TOLERANCE ONE WAY THAN ANOTHER—BIDIRECTIONAL

AS DRAWN

MEANING

NOTE Remaining part profile tolerance zones are determined by overall dimensions.

GAGE

NOTE Gage makers tol apply

DATUMS

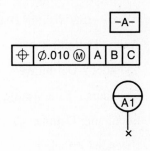

DATUMS

In the previous sections we have made continual reference to datums and have implied the importance of their use. Intelligent and effective application of the principles of geometric dimensioning and tolerancing depends to a great extent on a good understanding of the kinds of datums used, their definitions, proper selections, and the interpretations implied by the various uses.

The first portion of this section on datums covers the various kinds of datums, their definitions, establishment, relationship to each other and to measuring planes, and their meaning.

The last portion of this section will cover applications, selection, and further sample part examples.

214

DATUMS

A datum is a theoretically exact point, axis, or plane derived from the true geometric counterpart of a specified datum feature. A datum is the origin from which the location of geometric characteristics of features of a part are established. Datums are established by specified features or surfaces. Where orientation or position relationships are specified from a datum, the features involved are located with respect to this datum and not with respect to one another.

Every feature on a part can be considered a possible datum. That is, every feature shown on a drawing depicts a theoretically exact geometric shape as specified by the design requirements. However, a feature normally has no practical meaning as a datum unless it is actually used for some functional relationship between features. Thus a datum appearing on an engineering drawing can be considered to have a dual nature: it is (1) a ''construction'' datum, which is the geometrically exact representation of any part feature, and (2) a ''relationship'' datum, which is any feature used as a basis for a functional relationship with other features on the part. Since the datum concept is used to establish relationships, the ''relationship'' datum is the only type used on engineering drawings.

By the above definition, a datum on an engineering drawing is always assumed to be ''perfect.'' However, since perfect parts cannot be produced, a datum on a physically produced part is assumed to exist in the contact of the actual feature surface with precise manufacturing or inspection equipment such as machine tables, surface plates, gage pins, etc. These are called simulated datum planes, cylinders, etc., and, while not perfectly true, are usually of such high quality that they adequately *simulate* true references. This contact of the actual feature with precise equipment is also assumed to simulate functional contact with a mating part surface.

DATUM FEATURE (SURFACE HOLE, SLOT, DIAMETER, ETC.)

A *datum feature* is an actual feature of a part that is used to establish a *datum*. A *feature* is the general term applied to a physical portion of a part, such as a surface, hole, or slot.

By the definition of a DATUM, which is assumed to be theoretically exact, a ''datum surface'' or a ''datum feature'' is technically *not* a datum. These terms are, however, used in referring to the actual part feature or surface *from which* a datum is established when the feature or surface is used as a reference or contact with a tool, surface plate, or other checking device (the *simulated* datum). A ''datum surface'' or ''datum feature'' refers to the actual part feature or surface used to *establish* a datum. Since ''datum surfaces'' and ''datum features'' are actual things, they cannot be perfect, and thus they include all the real irregularities and inaccuracies of the surface or feature. (See examples under DATUM PLANE.)

DATUM POINT

A datum point is that which has position but no extent, such as the apex of a pyramid or cone, the center point of a sphere, or a reference point on a surface for functional, tooling, or gaging purposes.

The ''apex of a pyramid or cone'' and ''the center point of a sphere'' are considered *construction* datums, which are used to construct the geometry of the feature. ''A reference point on a surface for functional, tooling, or gaging purposes'' is considered a functional *relationship* datum and has meaning in the drawing specification. It may be implied or specified, dependent upon the extent of control desired.

An example of implied datum point and a specified datum point are shown following. *Implied* datum points are usually the random extremities or high points of an actual specified datum feature, which may be used to establish primary, secondary, and tertiary datum planes of orientation for the part. They occur at random locations as a part of the actually produced datum surfaces. Using the principles of geometry, we establish a ''primary'' datum plane by three (minimum) datum points, a ''secondary'' datum plane by two (minimum) datum points, and a ''tertiary,'' or third, datum plane by one (minimum) datum point.

Where datum planes are to be established for specific locations on a part surface, be it for design, or for manufacturing and inspection repeatability, they may be *specified* as datum target points, using datum target symbols, as shown. Location dimensions are required with *specified* datum target points.

DATUM POINTS

IMPLIED DATUM POINT

DATUM TARGET POINT ON SURFACE

.500

.500

A1 *

SPECIFIED DATUM POINT

*SPECIFIED DATUM TARGET. SEE FOLLOWING PAGES FOR EXPLANATION.

DATUM LINE, DATUM AXIS

A datum line is that which has length but not breadth or depth, such as the intersection line of two planes, the center line or axis of holes or cylinders, or a reference line for functional, tooling, or gaging purposes.

Examples showing an *implied* datum line and *specified* datum lines are shown following. Note that a hole (part a) contains a datum center line or axis (as every round hole does) which is derived from the intersection of two imaginary planes at an angle of 90°. The datum center line is also an axis as derived from the geometric center of a cylinder. These are construction datums inherent in the geometry.

However, unless the hole is *specified* as a datum feature (part c) with a relationship to other features, it has no meaning as a datum. When stated as datums, the same construction datums that are shown in part (a) become part of the requirement as established from the actual datum hole surface. (See also DATUM AXIS in following text.)

Where a specific datum line is to be established on a surface for design function, or manufacturing and inspection repeatability, it may be *specified* as a datum target line, as shown in part (b).

DATUM LINE, DATUM AXIS (continued)

IMPLIED DATUM CENTER LINE   SPECIFIED DATUM TARGET LINE   SECTION OF TWO PLANES)

(a)   **(b)**   **(c)**

*SPECIFIED DATUM TARGET. SEE FOLLOWING PAGES FOR EXPLANATION.

DATUM AXIS

A datum axis is a theoretically exact axis derived from the true geometric counterpart of a specified datum feature (i.e., a cylinder). A datum axis is the origin from which the relationship of geometric characteristics of features of a part are established. A datum axis is the theoretically exact center of the simulated datum cylinder as established by the extremities or contacting points of the actual datum feature cylindrical surface or the axis formed by the intersection of two datum planes. For an external feature, the datum is the axis of the smallest circumscribed cylinder which contacts the feature surface. For an internal feature, the datum is the axis of the largest inscribed cylinder which contacts the feature surface.

Since measurements or reference cannot be made from *theoretical* cylinders, etc., they are therefore assumed to exist, not in the part itself, but in the contact of the part with more precise manufacturing or inspection equipment. Fixture or gage pins, or gage cylinders, are *not* true cylinders, but they are usually of such high quality that they adequately simulate true cylinders, etc., and therefore are considered "true." These references are called *simulated* datum cylinders.

DATUM PLANE

A datum plane, when developed from a primary datum feature which is nominally flat, is a theoretically exact plane established by the extremities or contacting points of the *actual* feature surface with a *simulated* datum plane (surface plate or other checking device).

Since measurements or references cannot be made from *theoretical* planes, the planes are therefore assumed to exist, not in the part itself, but in the contact of the part with more precise manufacturing or inspection equipment. Machine tables, surface plates, fixture surfaces, etc. are *not* true planes, but they are usually of such high quality that they adequately simulate true planes and are therefore considered true references. These references are called *simulated* datum planes. Such reference planes can also be developed by coordinate movements of tooling or gaging equipment, such as on a coordinate measuring machine.

In the example below, the desired surface is shown as a datum feature and a datum plane. Also shown is the establishment of the datum plane from the actual part feature surface through contact of the surface extremities with the simulated datum plane. A relationship of this datum plane to other features on the part would be shown on the drawing.

DATUM PLANE ESTABLISHED FROM A
DATUM FEATURE (SURFACE)

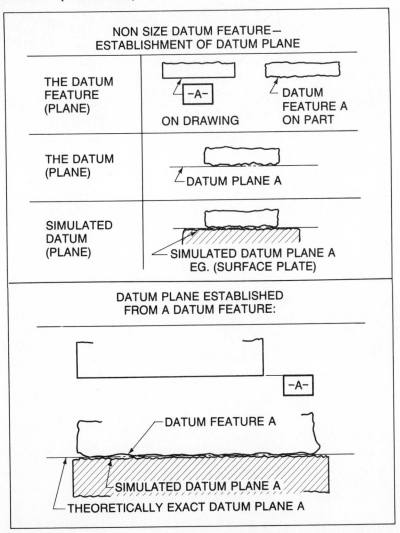

DATUM CENTER PLANE

A datum center plane is the theoretically exact center of the simulated datum. For an external width feature, the datum is the center plane between two parallel planes which, at *minimum* separation, contact the corresponding surfaces of the feature.

POSITION TOLERANCE TO DATUM CENTERPLANE, RFS

AS DRAWN

MEANING

For an internal feature, the datum is the center plane between two parallel planes which, at *maximum* separation, contact the corresponding surfaces of the feature.

POSITION TOLERANCE TO DATUM CENTERPLANE, RFS

ESTABLISHING DATUMS

ESTABLISHING DATUM PLANES FROM DATUM SURFACES— DATUM REFERENCE FRAME

Datum planes are theoretically perfect reference planes. Datum planes are always established by, or are relative to, actual or physical features. The most common datum plane is the type established from a datum surface.

The example under "DATUM PLANE" illustrated the conventional establishment of a primary datum plane from a primary datum surface. In establishing datum planes in relation to defining or measuring a part, at least two and usually three datum and measuring planes are considered in locating features. In part configuration, other than cylindrical, there are usually three planes of orientation. These three planes conform to the relationship of the conventional geometric X, Y, and Z axes and resulting planes of orientation.

The three planes are referred to as a datum reference frame and are composed of a primary, secondary, and tertiary (third) datum plane as established from the appropriate actual datum surfaces. Unless otherwise specified or controlled, the largest or most important surface is usually selected as the primary datum, the next largest or most important as the secondary datum, and the remaining surface as the tertiary plane. Design functional requirements should be the first criterion for the establishment of datum priorities. Where the datums are to be specified, they are identified with appropriate letters as previously discussed.

All datum planes contained within a datum reference frame are 90° BASIC, or perpendicular, to each other by interpretation. The dimension lines or center lines related to and/or shown perpendicular to these datum planes are implied as 90° BASIC, or perpendicular, to the datum planes. The resulting measuring planes are therefore considered to be mutually perpendicular to one another.

The illustration shows the establishment, orientation, and relationship of three datum planes from the part surfaces to determine the datum reference frame. The relationship of the measuring planes to the datum planes is also illustrated.

The PRIMARY datum surface A establishes the first relationship of the part for proper orientation. The three (minimum) extremities or contact points of the surface establish the primary datum *plane* A.

The SECONDARY datum surface B establishes a further relationship of the part for proper orientation. The two (minimum) extremities or contact points of the surface establish the secondary datum *plane* B at 90° BASIC to datum plane A.

The TERTIARY (third) datum surface C completes the part orientation. One (at least) extremity or contact point of the surface establishes the third datum *plane* C at 90° BASIC to both datum planes A and B.

Auxiliary datums, specified with respect to feature relationships not related to or functional with the part main datum reference frame, would not be considered 90° BASIC to the main datum system, but rather as 90° BASIC to the auxiliary datum reference frame.

ESTABLISHING DATUM PLANES FROM DATUM SURFACES/ FEATURES—THREE PLANE CONCEPT— DATUM REFERENCE FRAME

AS DRAWN

ESTABLISHING THE DATUM PLANES

ESTABLISH PRIMARY (FIRST) DATUM PLANE (3 POINT) (MIN) CONTACT WITH DATUM SURFACE A

ESTABLISH SECONDARY DATUM PLANE (2 POINT) (MIN) CONTACT WITH DATUM SURFACE B.

MEASURING DIRECTION FOR RELATED DIMENSIONS

ESTABLISH TERTIARY (THIRD) DATUM PLANE (1 POINT) (MIN) CONTACT WITH DATUM SURFACE C

ESTABLISHING DATUM PLANES FROM DATUM POINTS, LINES, OR AREAS (PARTIAL DATUM SURFACES) USING DATUM TARGETS

Where datum orientation is required on parts of irregular contour, such as castings, forgings, sheet metal, etc., datum targets provide a valuable tool. Specified datum targets which serve as means of constructing special datum planes of orientation can be of three types: points, lines, or areas. Datum targets establish the necessary datum reference frame and, in addition, ensure repeatable part location for manufacturing and inspection operations.

Datum targets are also used to indicate special, or more critical, design requirements where functional part feature relationships are to be indicated from specific points, lines, or areas on the part surface.

Datum points and lines have been previously defined. A datum area is a datum established from a partial datum surface. On a drawing, a datum area is outlined with phantom lines and identified by diagonal slash lines. It may be of any shape.

The previously mentioned geometric rules of establishing the datum planes of orientation (primary, secondary, tertiary) must be observed. However, where datum targets are used, particularly on irregular castings or forgings, the number of targets may vary from the conventional geometrical 3–2–1 point orientation as determined by the part configuration.

The locations and/or sizes of datum points, lines, or areas are controlled by BASIC or untoleranced dimensions and imply exactness within standard tooling, gaging, or shop tolerances. Toleranced locations or sizes may be used with datum target symbols if such practice is preferred.

The drawings below illustrate the use of datum target symbols to establish datum planes and part orientation.

DATUM TARGETS

PART ORIENTATION TO DATUM PLANES AS ESTABLISHED BY POINTS, LINES, OR AREAS

DATUM PLANES ESTABLISHED FROM DATUM POINTS AND AREAS ON NOMINALLY FLAT DATUM SURFACES:

AS DRAWN

INTERPRETATION

DATUM AREAS A1, A2, A3 (THREE CONTACTS) ESTABLISH PRIMARY DATUM PLANE A.

DATUM POINTS B1, B2, (TWO CONTACTS) ESTABLISH SECOND DATUM PLANE B ⊥ TO A

DATUM POINT C1 (ONE CONTACT) ESTABLISHES DATUM PLANE C ⊥ TO BOTH A AND B.

PART ORIENTATION FROM ESTABLISHED PLANES AS SHOWN

DATUM PLANES ESTABLISHED FROM DATUM POINTS AND LINES ON IRREGULAR CONTOUR SURFACES:

AS DRAWN

ESTABLISHING A DATUM AXIS FROM DATUM FEATURES

For a rectangular part, the three reference planes are readily visualized since the flat surfaces of the part resemble datum planes, and orientation to the three geometric planes X, Y, and Z is obvious. However, this is not true for cylindrical part surfaces since they bear no resemblance to datum planes and their relationship to the X, Y, and Z planes of orientation is not so obvious. Due to the geometry of a cylinder, a cylindrical part may be assumed to be simultaneously located in two planes of orientation. These two planes may be visualized as center planes intersecting 90° BASIC to each other at the datum axis. As discussed in the section *Datum Axis*, on outside cylinders the datum axis is established from the minimum circumscribed simulated datum cylinder contacting the actual cylindrical surface (RFS). On inside cylinders, it is established from the maximum inscribed simulated datum cylinder contacting the actual cylindrical surface (RFS).

Where the MMC principle is to be applied, the datum axis of a cylindrical part is established by the axis of the datum surface cylinder at MMC size. Under this condition, the actually produced feature axis may vary within a tolerance zone equal to the departure from MMC size of the feature. This situation is illustrated in the example, PRIMARY DATUM FEATURE MMC.

The illustrations following expand upon the application of datums to cylindrical parts. The examples and the illustrated interpretation of each should be self-explanatory. Note that a cylindrical feature, due to its geometry, establishes orientation in two directions simultaneously: the two imaginary datum construction planes intersect at the datum axis. However, only one datum letter identification is used to establish the datum feature (cylinder).

The examples also illustrate the use of a secondary datum plane with the primary datum feature (cylinder) and the use of the feature as *secondary* datum. In the latter instance, note that the secondary datum feature (cylinder) (perpendicular, 90° BASIC to the primary datum plane) may have its axis at variance with the *actual* feature axis. This possibility, however, is recognized and is a part of the design criterion whenever datums are used in this sequence on a part of cylindrical shape.

Where rotational orientation of the cylindrical part relative to features is required, a tertiary datum is necessary. The example, TERTIARY (THIRD) DATUM FEATURE RFS on pages 229 and 263, illustrates this type of requirement.

ESTABLISHING A DATUM AXIS FROM DATUM FEATURES

PRIMARY DATUM FEATURE RFS

PART

AND TO SECONDARY DATUM PLANE "B"

WITH RESPECT TO
PRIMARY DATUM "A", RFS

WITHIN .XXX FIM

THIS FEATURE TO BE IN CIRCULAR RUNOUT

MEANING

DATUM PLANE "B" PERPENDICULAR TO
DATUM "A" AXIS. ONE POINT
(MIN) CONTACT WITH DATUM
SURFACE "B".

TWO IMAGINARY
DATUM PLANES
INTERSECTING
AT AXIS OF "A".

TO DATUM "A" RFS
WITHIN .XXX FIM, RFS

DATUM "A" AXIS

DATUM
AXIS "A"

PRIMARY DATUM FEATURE "A"
IN CONTACT WITH SIMULATED DATUM
CYLINDER "A", THE SMALLEST CIRCUMSCRIBED
CYLINDER, ESTABLISHES DATUM AXIS "A".

MEASURING DIRECTIONS FOR
RELATED FEATURES

PRIMARY DATUM FEATURE MMC

PART

.XXX

Ø.XXX ±.XXX

-A-

| ⊕ | Ø.XXX Ⓜ | A Ⓜ | B |

AND SECONDARY DATUM PLANE "B"
WITH RESPECT TO PRIMARY
DATUM "A" AT MMC
WITHIN Ø.XXX AT MMC
THIS FEATURE TO BE LOCATED AT TRUE POSITION

4X Ø.XXX ±.XXX

| ⊕ | Ø.XXX Ⓜ | A Ⓜ | B |

.XXX ±.XXX

-B-

Ø.XXX ±.XXX

| ⊕ | Ø.XXX Ⓜ | A Ⓜ | B |

DATUM PLANE "B" ⊥ TO
DATUM "A" FEATURE AXIS AT MMC
ONE POINT (MIN) CONTACT WITH
DATUM SURFACE "B".

MEANING

Ø.XXX ⊕ TOL
ZONE AT FEATURE
MMC

POSSIBLE AXIS OF
DATUM "A" FEATURE
AWAY FROM MMC.

TWO IMAGINARY
DATUM PLANES
INTERSECTING
AT AXIS OF "A"
AT MMC SIZE

⊕ TO DATUM "A" AT MMC
WITHIN .XXX AT MMC SIZE
OF FEATURE

DATUM
AXIS

AXIS ZONE EQUAL IN SIZE
TO DIFFERENCE BETWEEN
ACTUAL SIZE AND MMC
SIZE OF DATUM "A" FEATURE

90°

90°

DATUM "A"
AXIS

PRIMARY DATUM
FEATURE "A"
TRUE CYLINDER
ESTABLISHED BY MMC
SIZE OF PRIMARY
DATUM FEATURE

DATUM "A" FEATURE
FEATURE
FINISHED AWAY
FROM MMC.

POSSIBLE AXIS OF
DATUM FEATURE
WHEN AWAY FROM MMC.

MEASURING DIRECTIONS FOR
RELATED FEATURES

SECONDARY DATUM FEATURE RFS

PART

Ø.XXX ±.XXX

| ⟋ | .XXX | A | B |

-A-

Ø.XXX ±.XXX

-B-

.XXX ±.XXX

| // | .XXX | A |

| ⟋ | .XXX | A | B |

TO DATUM AXIS
"B", RFS
WITH ORIENTATION FROM
PRIMARY DATUM PLANE "A"
WITHIN .XXX FIM
THIS FEATURE TO BE WITHIN
CIRCULAR RUNOUT

PRIMARY DATUM PLANE "A"
THREE POINT (MIN) CONTACT
WITH DATUM SURFACE "A."

MEANING

MEASURING DIRECTIONS FOR
ANY RELATED FEATURES

⟋ TO DATUM "B" RFS WITHIN
.XXX FIM, RFS

PRIMARY DATUM
PLANE "A"

DATUM
AXIS

TRUE CYLINDER
⊥ TO "A"

POSSIBLE AXIS
OF SECONDARY
DATUM FEATURE

DATUM AXIS

90°

90°

DATUM AXIS (AXIS OF TRUE CYLINDER
⊥ TO PRIMARY DATUM PLANE "A"
AND CONTACTING EXTREMITIES
OF SECONDARY DATUM FEATURE "B", RFS.)

TWO IMAGINARY DATUM PLANES
INTERSECTING AT DATUM AXIS

SECONDARY DATUM FEATURE MMC

PART

MEANING

AND TO SECONDARY DATUM CYLINDER B MMC
WITH ORIENTATION FROM PRIMARY DATUM PLANE A
WITHIN ⌀.XXX AT MMC
THIS FEATURE TO BE LOCATED AT TRUE POSITION

TERTIARY (THIRD) DATUM FEATURE RFS

PART

MEANING

ESTABLISHING DATUM CENTER PLANES FROM DATUM FEATURES—DATUM REFERENCE FRAME (THREE PLANE CONCEPT), DATUM PRECEDENCE

MEANING NO. 1

MEANING NO. 2

NOTE The "open set-up" interpretations above and on the preceding page, MEANING 1, demonstrate the principles involved. Production gaging methods, or special gages patterned after these techniques, can facilitate more effective inspection, e.g., equalizing mechanisms can be used in lieu of the measuring and shimming method (see MEANING NO. 2).

DATUM APPLICATIONS

The following section presents details of selection, application, and interpretation of datums under sample part conditions. Not all conditions are covered, but the examples shown are representative of a variety of typical datum applications.

The proper selection of datums is a very important aspect of geometric tolerancing. As previously emphasized, datums must be specified. In determining datums, we must consider first of all the part and feature relationships within the framework of the design requirements.

Where the relationships of the features are not of critical importance to part function or design intent, specific datums may not be required in the drawing specifications. However, usually some form of datum orientation for part manufacture and inspection is necessary. Therefore, whenever implied datums are invoked in the absence of specified engineering drawing datum detail, the designer depends on the discretion of the production and inspection departments.

Where specific relationships of features, datum precedence, and part orientation are of critical concern to design and manufacture, datums should always be specified. Of prime concern is the clarity with which the drawing conveys the design intent. Ambiguity should always be avoided, since it causes confusion and leads to misinterpretations.

Datum specification may appear to be a costly approach. Not so—on the contrary, it provides considerable advantages that pay off in economic production. It provides for clear instructions, protects the design intent, and ensures production and inspection follow-through in keeping with the intent.

The examples shown highlight the importance of adequate and proper datum specification. The illustrations on the left are not provided with datums, and the feature relationships are in doubt. The addition of datums in the examples on the right clarifies the relationships and thus ensures uniform interpretation.

SELECTION OF DATUMS

1. *Features* which are selected to establish *datums* must be clearly identified and/or easily recognizable. Datums must be specified, and must clearly represent necessary design intent.

2. *Corresponding features* on mating parts should be used in establishing datums to facilitate calculations and ensure proper part assembly.

3. To be useful for measuring, a datum on an actual piece should be accessible during manufacture so that *measurements* from it *can be made* readily.

4. *Avoid ambiguity* of datums by specifying datums where necessary for clarity.

SELECTION OF DATUMS

AS DRAWN

PERPENDICULAR TO WHAT? TOP OR BOTTOM SURFACE?

PERPENDICULAR TO DATUM SURFACE "A"; NOW CLEAR!

MEANING

POSITION OF HOLE PATTERN WITH RESPECT TO WHAT!
CENTER HOLE?
COUNTERBORE?
O.D.?

POSITION OF HOLE PATTERN WITH RESPECT TO THE COUNTERBORE, DATUM "A"; NOW CLEAR!

ACCURACY OF A DATUM SURFACE ON AN ACTUAL PIECE

The accuracy or quality of surfaces used for datums depends on the design requirement. The desired control, such as the form tolerance of flatness, may be specified where necessary.

Earlier U. S. standards have suggested the need to control datum surfaces to a finer degree of accuracy than that provided by dimensions or relationships taken *from* the datum surfaces. Often design requirements may call for this refinement and the need for it should therefore be considered. However, as an overall rule, a request for specific accuracy of the datum surface should be based upon the design requirements and the functional purpose of the surface.

By the definition of datum, a datum surface may have inaccuracies which will have no effect on the features located from the datum plane (as established from the extremities of that surface).

The illustration shows the two conditions of datum accuracy for features located from the datum, one where datum surface accuracy is *less* than that for feature location related to the datum (a), and one where the datum surface is held *more* accurate than the feature location related to the datum (b). Either requirement could be valid depending upon the design requirement.

Unless otherwise specified, part surface quality is normally a result of the type of machining process used to achieve size tolerances. Often, workmanship or local manufacturing tolerances will be the determining factors. Thus, where surface accuracy is of no critical concern, it need not be specified, yet that surface can serve as a datum reference from which other relationships can be specified.

DATUM SURFACE REQUIREMENT
LESS ACCURATE THAN LOCATION
TAKEN FROM DATUM PLANE

(a)

DATUM SURFACE HELD MORE
ACCURATE THAN LOCATION
TAKEN FROM DATUM PLANE

(b)

SPECIFIED DATUM SURFACES USING THE FORM TOLERANCE ⊥ AND POSITION ⊕ DIMENSIONING AND TOLERANCING

When feature interrelationships involving form, orientation, and position tolerances are to be more accurately and directly controlled, it becomes necessary to specify datums on the drawing so that such relationships can be clearly indicated.

In this example, the top surface of the part is identified as datum A, the left edge as datum B, and the lower edge as datum C. Once these identifications are made, the specific relationships between surfaces and features can be stated.

As previously described, datum planes are established from the extremities of the actual datum surfaces. Each relationship is thus taken from the datum plane. The datum surface inaccuracies, not coincidental with the datum plane, are not involved in the relationship.

The interpretation illustrates the establishment of the datum planes from the specified datum surfaces. The specified datums were used in this example to control orientation or perpendicularity of certain surfaces to precise limits. They were also used to give datum precedence orientation from the part surfaces to the position toleranced pattern.

AS DRAWN

3X ∅.250 ±.002

| ⊕ | ∅.010 Ⓜ | A | B | C |

.620

2.00 ±.01

.630

| ⊥ | .003 | A |
| ⊥ | .004 | B |

−C−

.500

.500

.500 ±.005

1.75 ±.01

−A−

| ⊥ | .002 | A |

−B−

MEANING

(**)
1.750

(a)

DATUM PLANE A

(b)

90°

.500

.500

DATUM SURFACE

(*) ACTUAL HOLE CENTER
WITHIN POSITION
TOL ZONE (3) .010 DIA
AT MMC SIZE .248.
.014 DIA AT HI LIMIT
SIZE .252.

90°

.620

(**)
2.000

90°

.630

DATUM PLANE C

DATUM
SURFACE

(**)
.500

DATUM PLANE B

DATUM
SURFACE

TOL ZONE
.004 (⊥ TO DATUM
PLANE B)

TOL ZONE
.003 (⊥ TO
DATUM PLANE A)

DATUM PLANE A

TOL ZONE .002
(⊥ TO DATUM
PLANE A)

90°

(c)

NOTE *Tolerance zones apply as normally measured from the datum planes. If
measured by some method where reference on the datum surface is not
coincidental with the datum plane, measurement is affected by the datum
surface inaccuracy (deviation between datum surface and datum plane) at that
point.
**Size measurement of individual feature per Rule 1.

The left datum surface is specified as perpendicular to datum plane A within .002. The lower datum surface is specified as perpendicular to datum plane A within .003 and to datum plane B within .004.

In view (b) in the lower interpretation the .003 wide perpendicularity tolerance zone is established between two parallel planes exactly 90° BASIC (or perpendicular) to datum plane A. The actual surface must lie within this .003 wide tolerance zone when it is measured relative to datum plane A anywhere along the entire length of its surface.

In (a), the .004 wide perpendicularity tolerance zone is established between two parallel planes exactly 90° BASIC (perpendicular) to datum plane B. The actual surface must be within this .004 wide tolerance zone when it is measured relative to datum plane B anywhere along the entire length of its surface. In (c), datum surface B is to be held perpendicular to datum A within a .002 width tolerance zone 90° BASIC to datum plane A.

Note that the measurements are taken from the datum planes which are theoretically perfect as established by the extremities of the actual datum surfaces. The perpendicularity tolerance zones establish the limit of the perpendicularity or form inaccuracies of these surfaces.

As in the note to view (a) at left, the actual hole centers must fall within the position tolerance zone of ϕ .010 when the hole is at MMC size of .248. The tolerance zone can increase to ϕ .014 if the holes are produced to the high limit (LMC) size of .252. The position tolerance includes both orientation (perpendicularity) and position error control.

The position tolerance zones are basically parallel to each other and the center planes of the pattern are basically perpendicular and parallel to each other. In addition, the position tolerance zones are basically perpendicular to the primary datum plane A, and parallel to the secondary and tertiary datum planes B and C.

The edge surface tolerance zones illustrated relative to the size dimensions 1.750 and 2.000 (** in the preceding figure) are intended to relate the edge extremities to the total part requirements. However, note that the interrelationship (perpendicularity) of these edge surfaces to each other is not controlled by the stated requirements; instead, it is subject to the considerations of Rule 1. The edge tolerance zones shown are representative of the size tolerance relationship of the individual 1.75 ± .01 and 2.00 ± .01 sizes only. Control of these surfaces in the relationship to one another would require specified orientation tolerances (e.g., perpendicularity), or it may depend on local workmanship or shop tolerances.

SPECIFIED DATUM SURFACES USING FORM TOLERANCES ⊥, ☐, AND POSITION ⊕ DIMENSIONING AND TOLERANCING

In the example on the facing page, the stated requirements and interpretation are identical to those in the preceding one except for the added flatness control of each of the datum surfaces. The accuracy or quality of the datum surfaces is specified.

Note that the flatness control is a refinement of other controls and is contained *within* the other orientation tolerance zones. Where combined form and orientation tolerances are applied, the predominant control (in this case perpendicularity) confines the tolerance latitude of additional controls (such as flatness) as a further refinement.

Datum planes are established from the datum surfaces as previously explained; the orientation to these planes is as shown in the preceding example.

AS DRAWN

MEANING

NOTE *Tolerance zones apply as normally measured from the datum planes. If measured by some method where reference on the datum surface is not coincidental with the datum plane, measurement is affected by the datum surface inaccuracy (deviation between datum surface and datum plane) at that point.

**Edge tolerance zone interrelationship to each other, subject to considerations of Rule 1, size measurement of individual feature within boundary of perfect form at MMC.

SPECIFIED DATUM FEATURES (SURFACE, HOLES) USING FORM TOLERANCES ⊥, ▱, //, AND POSITION ⊕ DIMENSIONING AND TOLERANCING

In this example (Flange Mount) we use a variety of geometric controls. Note the manner in which datums are applied so that direct relationships may be controlled per the part function as seen in the lower illustration.

The large base surface in Fig. (a) is to be held flat within .001 and is also identified as datum plane A. Note that the ⌀ .750 center hole diameter is to be perpendicular to datum plane A within ⌀ .0015 and is also established as datum B.

Also in Fig. (a), the ⌀ 1.250 diameter is to be held to a positional tolerance of ⌀ .004 at MMC with respect to datum B at MMC; the 1.4375–16 UN–2A thread is to be held to a positional tolerance of ⌀ .004 at MMC with respect to datum B at MMC; the face adjacent to the thread undercut is to be parallel within .003 with respect to datum A; and the ⌀ 1.750 counterbored hole at left is to be held in total runout within .001 FIM with respect to datum B and is also established as datum C.

The six holes shown in Fig. (b) are to be located at true position within ⌀ .005 at MMC with respect to datum plane A and datum C at MMC. Datum plane A identifies the surface from which the positional pattern is related and orients the hole center tolerance zones as basically perpendicular in attitude to this plane. The theoretical center of datum hole C establishes the center of the 3.375 BASIC diameter and the 60° BASIC angles.

The lower Fig. (c) illustrates the part function (the Flange Mount) as it relates to the mating parts in assembly. From this assembly requirement the critical nature of the perpendicularity, runout and position (mounting holes) can be seen. The interrelationship of all the related features is controlled by the appropriate geometric tolerances and datum references.

AS DRAWN (FLANGE)

(a)

(b)

ASSEMBLY

HOUSING

SHAFT

FLANGE MOUNT

(c)

SPECIFIED DATUM POINTS ON SURFACES USING POSITION $\oplus$ DIMENSIONING AND TOLERANCING

This example illustrates a part whose function requires that specific locating or reference points be established on the edge surfaces. These points are *datum* points and are identified by datum target symbols.

Usually, surface datum points are used either to simulate the desired location of contact with the mating parts and thus hold the related dimensions and tolerances with respect to these points, or to assure repeatable location and orientation of the part for successive fabrication and inspection operations on castings, forgings, or sheet metal parts that have irregular surfaces.

Note that the datum points are dimensioned with BASIC or untoleranced dimensions. By interpretation, BASIC dimensions are theoretically exact and, in this case, represent the ideal location for which to strive in order to relate the holes with the part edges. In these circumstances and in actual applications, the BASIC dimensions are assumed to represent exactness within tooling, gaging, or shop tolerances. Where necessary, toleranced locations or sizes may be used with datum target symbols.

Actually, as previously explained, the datum points establish datum planes. Thus, the position pattern is simultaneously related to these points and the planes as established from these points.

The two holes on the left, the ϕ .300 ± .002 holes, are to be located at true position within ϕ .006 at MMC with respect to datum planes A, B, and C. The third hole at the right, the ϕ .130$^{+.002}_{-.000}$ hole, which also is a part of the hole pattern, is to be located at true position within ϕ .004 at MMC with respect to datum planes A, B, and C.

The meaning illustrates the relationship of the hole pattern to the datum points and established planes.

Datum points B1 and B2 in the left view establish a datum plane B at 90° BASIC to the top datum plane A. From this plane B, the vertical dimensions are taken. Thus, these dimensions are related *only* to points B1 and B2 and their common plane and *not* to the entire surface.

Datum point C1 establishes a third plane, C, 90° BASIC to both datum planes A and B. Only one point is, of course, necessary on the third plane. The horizontal dimensions are related *only* to point C1 and its common plane, C, and *not* to the entire surface.

Note that the positionally toleranced pattern of holes is oriented with respect to the established datum planes, since, by interpretation, all center planes are basically parallel and perpendicular to the edge datum planes.

The actually produced hole centers are permitted to vary within the tolerance zones stated on the drawing. However, as the positional relationships are specified at MMC, the positional tolerance of the two larger holes may possibly increase from ϕ .006 (at MMC size of .298) to ϕ .010 when the holes are at LMC size, or largest size, of .302. The positional tolerance of the third hole may possibly also increase from ϕ .004 (at MMC size of .130) to ϕ .006 at LMC size, or largest size, or .132.

AS DRAWN

MEANING

(*)
HOLE TOL ZONE
.004 DIA AT MMC (.130)
.006 DIA AT MIN MC (.132)

(*)
HOLE (2)
TOL ZONE
.006 DIA AT MMC (.298)
.010 DIA AT MIN MMC (.302)

TARGET C_1
SECONDARY
DATUM PLANE
ESTAB. BY B1, B2

B_1 TARGETS B_2

DATUM
PLANE A

TARGET
C1

TERTIARY
DATUM PLANE
ESTAB. BY C1

DATUM SURFACE
A

PRIMARY
DATUM PLANE A

TARGETS
B_1
B_2

NOTE   *Tolerance zones apply as measured from the datum planes which have been
established by the datum targets.
**Edge tolerance zone interrelationship to each other, subject to considerations
of Rule 1, size measurement of individual feature within boundary of perfect
form at MMC.

SPECIFIED DATUM POINTS ON SURFACES USING POSITION ⊕ DIMENSIONING AND TOLERANCING

Note again that when using this form of position dimensioning and tolerancing with datum points, the position pattern and thus the individual holes are related to the datum points and their established planes only—they are *not* related to the entire surface.

The deviations or imperfections of the remainder of the actual surfaces on which the specified datum points are established are controlled by the overall coordinate dimensions. See note (**) at lower part of illustration. The edge surface tolerance zones illustrated are intended to relate the edge extremities to the total part requirements. However, note that the relationship (perpendicularity) of the edge surfaces to each other is *not* controlled by the stated requirements. This interrelationship is subject to Rule 1. The edge tolerance zones shown are representative of the size tolerance relationship of the individual 2.00 ± .01 and the 2.25 ± .01 sizes only. Control of these surfaces in relationship to one another would require specified form or attitude tolerances (e.g., perpendicularity), or may depend on local workmanship or shop tolerances.

Typical applications of this form of dimensioning and tolerancing are an irregular surface, such as a casting, which requires specific tooling or gaging pickup for repeatability, or a part whose function requires specified location of points of contact or relationship with a mating part. When we locate these points and identify them, we ensure that specific design or tooling and gaging requirements are clearly stated and that the manufacturing and inspection processes will follow well-defined instructions in agreement with the part requirements.

Usually, in an application such as this, the datum points would be picked up by locating buttons. There is no implication that detents are to be applied to the part surface. If detents are desired, a note to this effect should be added to the drawing.

If positional dimensioning located by BASIC dimensions from an edge surface is employed and *no* datum points nor planes are specified, the high points or extremities of the implied datum surface will be used to establish datum planes and orient the position pattern of features to the edges. In this case, we would use conventional *flat* locating surfaces instead of buttons. Although the resulting control of the relationship of the position pattern of features to specific portions of the part edges would be less stringent, it may be adequate in many applications.

AS DRAWN

MEANING

NOTE   *Tolerance zones apply as measured from the datum planes, which have been established by the datum targets.

**Edge tolerance zone interrelationship to each other, subject to considerations of Rule 1, size measurement of individual feature within boundary of perfect form at MMC.

SPECIFIED DATUM LINES ON SURFACES

In this example, we expand the use of datum points to establish datum lines on surfaces. We see that selecting specific datum or reference lines as the basis for establishing relationships to other features often has the advantage of providing effective and economical methods of achieving design intent or part function.

This illustration shows a rotor. It is not completely dimensioned—only enough to explain how the datums are applied.

Datum lines A1 and A2 specify the location of two specific lines on the rotor shaft which are to be the basis for the runout relationship with the rotor outside diameter.

Datum lines A1 and A2, because they are points on a diameter, actually apply to an infinite number of points around the circumference of the shaft to establish the datum lines. As was previously explained with respect to the BASIC dimensions, exactness within tooling, gaging, or shop tolerances is implied. Where necessary, toleranced locations or sizes may be used with datum target symbols.

In actual applications, the gaging pickup of the datum lines would, for example, be accomplished by two vee blocks of very narrow width. The datum lines essentially simulate the location of the bearing races and the accuracy required at these lines. This eliminates the need for holding the critical runout requirement on a larger portion of the shaft length than is necessary to ensure proper function of the part. The remainder of the shaft is then controlled by shop tolerances or other specifications.

The runout notation would be interpreted to mean "the rotor diameter is to be within .001 FIM total runout with respect to datums A1 and A2 (and their common datum cylinder and axis) simultaneously."

AS DRAWN

MEANING

DATUM LINES A₁ & A₂ CREATE DATUM AXIS A.

1.7500 DIA TO BE WITHIN TOTAL RUNOUT OF
.001 FIM WHEN MOUNTED ON LINES A₁ & A₂.

* = OR COMPARABLE HOLDING DEVICE

SPECIFIED DATUM POINTS AND LINES ON SURFACES

Figure 1 shows a selection of two datum points, A1 and A2, used to specify a relationship of parallelism between two points on the part surface and the elongated slot.

FIGURE 1

Figure 1 shows that these two datum points, A1 and A2, or more specifically, the theoretically exact datum line A extending between these two datum points (and not the entire surface) establishing the datum reference from which the parallelism of the slot is to be established.

The surface upon which datum points A1 and A2 are established could have curvature or inaccuracies which would have no effect on the parallelism requirement. The accuracy of the entire surface is controlled by the vertical $1.130 \pm .005$ dimension. The parallelism requirement is specified to be from the common datum line A established by datum points A1 and A2 to the center line of the slot within .002 total.

Figure 2 shows datum line A1 (at the left) established on the angular surface of the part. Datum line A1 is used in conjunction with the center of datum feature B to establish a center plane which determines the reference from which the 30° BASIC angle to the slot is taken.

FIGURE 2

SPECIFIED DATUM AREAS ON SURFACES

FIGURE 3

If the datum reference must be taken from given small areas on the primary datum surface (Fig. 3), the desired datum surface area may be shown as a diameter (or other shape), with location and size of the area controlled by BASIC dimensions. The datum surface area limits are enclosed (with phantom-line circles in this case) and the area is identified by diagonal slash lines.

The extremities of the three datum surface areas A1, A2, and A3 combine to form a single, or common, datum plane A.

Three .250 flat-ended buttons or locators contacting the surface at the prescribed locations will provide the actual location. The other requirements are then related only to these datum surface areas and their common resulting plane and not to the entire surface.

SPECIFIED DATUM AREAS ON SURFACES

In some applications, it is convenient to select larger areas of surfaces as datums.

In Fig. 1, portions of the shaft diameter identified as datums A1 and A2 are the basis for the runout relationship to the shaft step diameter.

FIGURE 1

As in the previous example on datum points, the datum reference applies to the circumference around the shaft contained within the limits of the basic dimensions. The extent of the datum surfaces is the .500 BASIC and .400 BASIC wide reference bands or areas on the shaft diameter. These datum surface areas simulate the location of bearing races.

It can be seen that the shaft step diameter of $\emptyset 1.2500 \pm .0005$ is required to be held within total runout of .001 with respect to the two datum area diameters A1 and A2 and their established common datum reference axis A. The remainder of the shaft is permitted to deviate within more lenient standard or shop tolerances.

As in the previous examples, exactness within tooling, gaging, or shop tolerances with respect to the basic dimensions is implied.

Figure 2 illustrates a metal blade. The $1.400 \pm .010$ dimension is to be held between the lower flat surface and the contact button. However, in addition, there is a .005 total parallelism requirement from that portion of the lower surface identified as datum A1 and the button.

The limitations of datum target A and its established datum reference plane are specified by a phantom line and a BASIC dimension, with the area identified by diagonal slash lines drawn across the surface. This clearly shows that the surface so identified determines the datum reference plane.

FIGURE 2

Figure 3 is another example in which a portion or area of a surface serves as a datum reference. This application was based on the part function and the anticipated difficulty of holding accuracy on any more of the formed part flat surface than necessary.

FIGURE 3

Note the manner in which the basic dimensions have been used to define the limits of the datum area A1. Area A1 establishes the datum reference plane A from which the measurement is taken.

The formed ears of the part are to be held parallel within .003 to the datum plane A.

Again, exactness within tooling, gaging, or shop tolerances is implied by the BASIC dimensions.

SPECIFIED MULTIPLE DATUM AREAS

In this example, three mounting bosses of a cast part are selected as datum target areas. The purpose is to clearly specify that the three bosses *only* are to establish datum plane A for the relationship of the parallelism requirement on the top surface. The extremities of the three individual boss surfaces would combine to provide the three-point (minimum) contact, establishing datum plane A. Normally complete features (i.e. the bosses) are *not* used as datum targets. However, in this case, the three bosses are used to compose *one* datum plane as the design requirement and plane of reference for the parallelism relationship. Usually datum target are *portions* of surfaces or features.

The top surface of the part is to be parallel within .003 to the established datum plane A. This guarantees the assembled part relationship of the parallelism requirement to the mounting bosses.

SPECIFIED "STEP" DATUMS

When we establish datum planes of orientation on parts of irregular shape such as a casting, the datum target points (and thus their datum pin locators) may involve surfaces which are at different levels and must therefore be *combined* to establish the desired plane of orientation.

Referring to the illustration on the next page, we see that datum points C1 and C2 are situated at different surfaces and at different levels (steps) in order to relate the entire bulk of the part to a common plane of orientation. This plane, identified as the common plane C, may then be related to the primary datum B and the tertiary datum D with the usual implied mutually perpendicular datum plane relationship. The step basic dimension is to be maintained within tooling or gaging tolerance accuracy.

Other features of the part, such as the position toleranced pattern, can then be related to the datum system as established from these "step" datums and the other interrelated datums to ensure compliance with design requirements and repeatability of manufacturing and inspection operations.

SPECIFIED "EQUALIZING" DATUMS

Page 253 illustrates a part on which the orientation is established from specified datum targets (lines and points) as a means of "equalizing" the bulk of the part to centralize the location of the three holes. Through the datum target method as specified, a guarantee of design requirement fulfillment is provided, and tooling and gaging repeatability is assured. The two planes of orientation necessary to this part are provided by the equalizing and centralizing placement of the datum targets. The specified position relationships controlling the hole locations are then established from the datums A, B.

The primary datum plane A is established from points A1, A2, and A3 at the two levels of the part and relative to the step difference. These points and the step difference combine to give the common primary datum plane of orientation.

The secondary datum plane B is established by the equalizing action of the fixed and movable vee's as they contact tangent to the part at an angle of 45° on the left end and at the indicated lines of contact on the right end. Datum center plane B is perpendicular to datum plane A. Some slight "cam-down" or hold-down of the part may be necessary in actual applications.

The effect of any "step" tolerance as it relates to the features located from this datum should be considered.

SPECIFIED "STEP" DATUMS

AS DRAWN

4X .250–28 UNF–2B ▽ .40 ±.01

| ⊕ | Ø.010 Ⓜ | B | C | D |

.250 Ⓟ

.250

1.250

1.750

.980

.880

1.620

D1

-D-

-C-

C1

C2

3.000

1.375

1.000

D1

// | .002 | A

-B-

-A-

C1

C2

MEANING

HOLE PATTERN
ORIENTED TO
DATUMS B, C, D

POINT
D1

1.620*

DATUM PLANE C
(REPRESENTS A COMMON
PLANE OF ORIENTATION FOR
THE PART AS ESTABLISHED BY
POINTS C1 & C2 AND THEIR
STEP DIFFERENCE)

PRIMARY DATUM PLANE B.
CONTACT WITH THREE OR
MORE EXTREMITIES

1.250*

90°*

POINT C1

POINT C2

3.000*

1.375*

1.000*

POINT
D1

*DIMS MUTUALLY ⊥ , // , & ON LOCATION WITHIN
STANDARD OR ESTABLISHED TOOLING OR GAGING
TOLERANCES UNLESS OTHERWISE SPECIFIED.

SPECIFIED "EQUALIZING" DATUMS

AS DRAWN

MEANING

(REPRESENTS A COMMON PLANE OF ORIENTATION FOR THE PART AS ESTABLISHED BY POINTS A1, A2 & A3 AND THEIR STEP DIFFERENCE)

*DIMS. WITHIN STANDARD OR ESTABLISHED TOOLING OR GAGING TOLERANCES.

EXTENDED DATUM PRINCIPLES

MULTIPLE DATUM REFERENCE FRAMES WITH NONCOMMON DATUMS

In multiple datum reference frames with noncommon datum references established on the same part to control specific part feature relationships, *no* specific relationship may exist between the datum systems. They are treated separately.

AS DRAWN

MEANING

FIRST DATUM REFERENCE FRAME

SECOND DATUM REFERENCE FRAME

INTERRELATED DATUM REFERENCE FRAMES—WITH COMMON DATUMS

Datum references, either single or double datums, or complete three-plane datum reference frames, which are interrelated on the same part with other datum reference frames, retain a relationship with the other reference frames to the extent of common datum references.

In some cases, one datum reference frame established within or related to another through common datum references involves refinement or extension of the other datum reference frame, and, if necessary for orientation, may retain a relationship with the other datum reference frame as implied by the drawing.

There are two position toleranced hole patterns on this part. The first one involves a single hole which is established relative to the datum reference frame A, B, C. This hole then becomes the secondary datum for the interrelated second datum reference frame A, D, B to which the second position pattern is related.

AS DRAWN

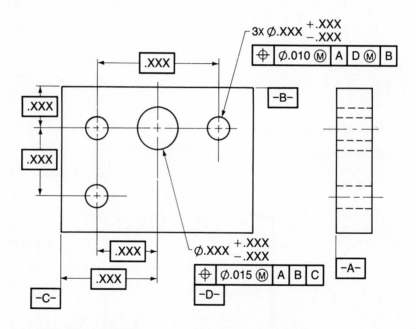

Datum D is directly related to the main datum reference frame (as established by A–B–C) by specification. The three-hole pattern is then related to datums A, D, and B. Datum A is repeated for clarification to ensure correct selection of the primary datum.

Since the establishment of the interrelated datum reference frame A–D–B within the main datum reference frame A–B–C is to achieve a refinement of the relationship between the three small holes and the large hole (datum D), the positional relationship is to be between these features only as specified. Orientation (or rotation) is controlled by datum B as the tertiary datum.

MEANING

FIRST REQUIREMENT (FIRST RELATIONSHIP—TO DATUM REFERENCE FRAME A, B, C)

SECOND REQUIREMENT
(SECOND RELATIONSHIP—TO INTERRELATED DATUM REFERENCE FRAME A, D & B)

INTERRELATED DATUM REFERENCE FRAMES—WITH COMMON DATUMS—TWO HOLES AS SECONDARY AND TERTIARY DATUMS

USE OF THE TWO DATUM FEATURE (HOLE OR PIN) PRINCIPLES

The use of two holes (or pins) to establish the key interface between mating parts is classic in designing and building hardware. Along with establishment of a primary datum feature (surface) as shown in the example at right and others in this section, the two size feature (holes) datums completely stabilize the part to the three planes. Occasionally, due to some unique design requirements, more than two features (or even a pattern) can be used as datums (see below and pages 126, 127). Nevertheless, no more than two size features are needed to provide complete orientation to the part. Where more than two features are used as datums, a mutual establishment of the two X and Y planes is derived via a centroid of rotation. Although feasible, and on occasion needed, this approach should be avoided if possible due to complications in interpretation and verification. Only two features are necessary to stabilize the part. Therefore, when a pattern is used as *one* datum, the question as to which two features actually do the locating is left to *any* two. It is seldom discernible (or necessary) to know which two. Yet in verification, unless some functional gage simulation is used, the selection of the appropriate features for orientation will vary by random choice.

COMPOSITE POSITION TOLERANCING—HOLE PATTERN AS SECONDARY DATUM

AS DRAWN

The design requirements determine the manner and precision needed to define the part details. Obviously, its producibility is also of concern. Therefore, some tempering of the design in consideration of production (with design approval) is normal. However, designing around manufacturing processes, unless pertinent to defining the engineering design requirements, is contrary to recommended design practices and the authoritative standards. The design requirement should predominate.

The *two hole* (or pin) datum application is an extension of the *one hole* (or pin) application covered on pages 128–137 and 258, 259. The figure from page 129 is repeated below for information. One hole (or pin) as a secondary datum feature establishes a secondary datum axis. However, the x and y planes need to be oriented (rotated) to another feature (datum surface B was selected) as a tertiary datum. This ensures that the hole pattern (.380 holes) are placed properly relative to the desired datum reference frame.

ONE HOLE AS SECONDARY DATUM

AS DRAWN

Following discussion extends the preceding coverage to *two holes* used as secondary and tertiary datums. The design requirement is different than the preceding, but the need to stabilize the part to the datum reference frame is the same. In this case, however, it is done with a datum surface and *two datum* features (holes).

In this section the two feature (hole) application is explored in some detail. It does not exhaust all possibilities; only those most common and typical to two part interface of this variety are covered. The dimensions and tolerances (size and position) selected here are, of course, hypothetical and are only to represent principles and methodology possible in such situations. The magnitude of the values would be established by the real design requirements and the results of calculations and other determinations by the designer. A theme of this section, however, is to address the considerations and "facts" that exist when such requirements are applied. Where a pattern of features must be closely related to "size" datums (i.e., D and E in this section) and must "follow" them, due consideration must be given to the resulting effects of all the features involved (i.e., The Datum Virtual Condition Rule, etc.). The various methods shown progress through different possibilities where the effect of the datum feature size and position is to be considered, and then proceed on to methods where it can be neutralized to some extent. The choice is based upon the best judgment for the design at hand. An example of a typical two hole secondary and tertiary datum application is shown. Further detail on this example and others follow in this section.

AS DRAWN

Another aspect of the theme of this section is that some precision is inherent in the subject type application. That is, if in fact a feature, or a pattern of features, is to "follow" other features (datums), a closer relationship must be assured by the tolerances applied. First, some needed precision must be given to the datum features themselves. Then it is seen that improved (greater) precision, or a refinement of some kind, must be specified from the datum features to the pattern or features "following" them. This is needed to ensure the *closer* relationship. The examples shown gravitate toward that which would probably be called rather close precision. This was chosen to represent better the objective of this method. Of course, the values chosen could vary toward a more lenient limit as based upon part function, size, and the relative precision required. The principles remain the same.

FUNCTIONAL GAGING APPLICATION AND ALTERNATIVES

Functional gaging possibilities and options are shown in the examples in this section to explain principles and methodology. A functional gage is in essence the "reconstruction" of the mating part from the requirements indicated on the design. It describes, as well, a representative mating part (or mating situation) which simulates the two parts in assembly. It also represents a "worst case" part which remains an acceptable part. The functional gaging approach is not required on such parts; it is an available option.

A functional gage would never accept a "bad" part but conversely could reject a border line "good" part. This is because in the standard method of allocating gage making tolerances, "some" part tolerance is utilized for the gage.

Functional gaging principles are shown in this section, and elsewhere in this text, for two primary purposes. One, this assists in the understanding of the principles involved where such methods can be used, i.e., when the MMC principle is valid and invoked. Reconstructing the part requirements (e.g., a hole pattern) from the engineering design into a reverse image (a gage) provides a quantified picture of the result. Two, functional gaging is a valid production tool and technique which has numerous technical and economic advantages. See later portions of this section for further details on functional gages.

Functional gaging principles can also be utilized *without* a functional gage. Alternative methods such as graphic analysis, paper gaging, "scatter-grams," etc., using various tools, including the computer, are at our disposal. For example, the results of a coordinately measured part (such as hole patterns shown in this section) using a coordinate measuring machine (CMM) or comparable method can be used to simulate functional gaging. Further, a *mathematical solution* from data derived from a CMM operation can be determined with the assistance of calculators and computer programs.

Therefore, functional gaging principles can be achieved via three different, but complementary, ways: (1) a functional gage, (2) physical graphic analysis (e.g., paper gaging), or (3) mathematically using a calculator or computer programs. Establishment of the computer programs, however, requires a superior knowledge of the technical principles involved and herein described.

See section on Graphic Analysis and Mathematical Analysis of Coordinately Measured Features (pages 281–289) in this text for further discussion of such techniques.

The graphic or mathematical methods may be necessary where the precision of the part may not permit functional gaging (insufficient tolerance can be derived for the gage build), where parts are rejected by a functional gage and are suspected as borderline good parts, where RFS specifications to the features controlled prevent use of a functional gage, where the functional gaging is not justified, etc. Further, it should be noted that the mathematical (calculator/computer) methods may be used to bypass functional gaging and graphic analysis completely, usually with greater accuracy as well as more rapidly.

TWO HOLES AS SECONDARY AND TERTIARY DATUMS AT MMC

There are two positionally toleranced hole patterns on this part. The first pattern is related to a complete datum reference frame established by A, B, C. The two holes established with the first datum reference frame are then used with the common datum A to become the secondary and tertiary datums for the second complete datum reference frame A, D, E.

Note that this is an example illustrating the rather common requirement of pickup of *two* reference holes from which other features are to be related. See also further discussion on following pages using two holes (or pins) as secondary and tertiary datums.

AS DRAWN

MEANING

FIRST REQUIREMENT (FIRST RELATIONSHIP – TO DATUM REFERENCE FRAME A-B-C)

MEANING

TERTIARY DATUM PLANE C

90°

SECONDARY DATUM PLANE B

.XXX

.XXX

90°

Ø.003 TOL ZONE AT MMC

.XXX

PRIMARY DATUM PLANE A

GAGE

GAGE PINS Ø.247 (.250 MMC MINUS .003)

SECOND REQUIREMENT (SECOND RELATIONSHIP – TO INTERRELATED DATUM REFERENCE FRAME A, D, E)

Ø.005 TOL ZONE AT MMC (5)

SECONDARY DATUM D

90°(5)

.XXX
.XXX
.XXX
.XXX

.XXX

.XXX

TERTIARY DATUM E

PRIMARY DATUM PLANE A

GAGE

Ø.247 GAGE PIN (VIRTUAL CONDITION)* (2)

DATUM D PIN

Ø.200 GAGE PINS (.205 MMC MINUS .005)

Ø.425 GAGE PIN (.430 MMC MINUS .005)

DATUM E PIN

*PER DATUM VIRTUAL CONDITION RULE

The similarity of the above situation to the simpler one shown here on a cylindrical part may serve to clarify this principle of orientation through the secondary and tertiary datums.

Note that the secondary datum cylinder D on this cylindrical part could be compared to the secondary datum hole (cylinder) D on the above flat part and that datum hole E on both parts give orientation (rotation control) as the tertiary datum. A complete datum reference frame is established in both parts to which other features on the parts can be related.

.XXX

.XXX

-D-   -A-

.XXX

.XXX

Ø.XXX ±.XXX

Ø.250 +.003 -.000

3x Ø.205 +.003 -.000

⊥ | Ø.005 Ⓜ | A

⊕ | Ø.003 Ⓜ | A | D Ⓜ

-E-

⊕ | Ø.005 Ⓜ | A | D Ⓜ | E Ⓜ

Where two holes (or pins) are to be used as datum locaters for mating parts, numerous possibilities and considerations could be taken into account. This section addresses the major options available and explores in some detail such applications.

First, earlier coverage in this text (pages 128, etc.) introduced the use of a "size" feature (hole) as a datum feature. When only *one* hole (or pin) is used as a datum feature, an orientation (rotational) datum is required to stabilize the X, Y (and Z) planes and their relationship to the features (other holes) taken from these datum planes. A surface feature is used in the cited examples as a tertiary datum feature for orientation (rotation), not location. This is to ensure the clarity of the design requirement and for uniformity of interpretation in production and verification. The datum reference frame, in this case, is established by a plane surface (primary datum), a hole (secondary datum), and a plane surface (tertiary datum).

Use of *two* holes as secondary and tertiary datum (see Fig. (a) on page opposite and preceding) provides the location *and* orientation (rotation) for the related features in the X, Y and Z planes. In this case, the datum reference frame is established by a plane surface (primary datum A), a hole (secondary datum D), and a second hole (tertiary datum E).

The functional interface on this part with its mating part is via the two ϕ .250 holes and the corresponding mating part pins. Thus, selection of these two features as datum features (D and E) for the other related features (ϕ .430 and ϕ .205 holes) seems logical. The related holes, the ϕ .430 and four ϕ .205 holes, are to "follow" the datum locating holes as functional to the interface with corresponding features on the mating part. MMC is used on the datum features (D and E) as desirable to the design requirement.

The positional tolerance (ϕ .003 at MMC) for the two ϕ .250 datums D and E features is determined by the usual calculation relative to the corresponding mating part pins. Note on the representative functional gage (Fig. (b) and preceding page), that the gage pin sizes are ϕ .247; virtual condition of the hole. This gage could verify the part locating holes as in their acceptable location and with respect to datums A, B and C.

Note that the second functional gage (Fig. (c) and preceding page) to verify the five hole pattern (ϕ .430 and four ϕ .205) also requires that the "pickup" of the datums D and E holes be by *virtual condition* pins, ϕ .247. This is per the Datum Virtual Condition Rule.

Since the datum "pickup" pins on *both* gages (Figs. (b) and (c)) are at virtual condition size (ϕ .247), the question can be raised as to how much difference (if any) there is between the use of features D and E as datums or not. Could the entire pattern (all seven holes) have been related as *one* pattern relative to datums A, B and C and have the same control essentially? Could the two gages have been combined as one and represent the design requirement and part function?

See continued discussion on following pages.

AS DRAWN

(a)

GAGE

(b)

(c)

Figure (a) on the opposite page suggests the option of the preceding page, i.e., all seven holes are related to the *one* datum reference frame A, B and C (if the B and C datums are inserted instead of the D and E datums). This would be a legitimate requirement, but the desired closer relationship between the ⌀ .430 and ⌀ .205 holes and the two ⌀ .250 holes (as datums) originally intended is no longer specified nor achieved.

From this discussion it can be derived that if the *closer* relationship between the five holes to the datum features D and E is, in fact, desired, further controls may be necessary. Also, from the discussion on the preceding pages, the representative gages reveal the narrow difference between the *with*, or *without*, datums D and E at MMC applications. Where open set-up, coordinate measuring machine (CMM), or other manual methods of verification are implemented, the MMC effect and virtual condition basis could inadvertently be overlooked by common misunderstanding; whereby, the relationship to datums D and E may be incorrectly assumed as on an RFS basis and, thus, would invoke a *closer* relationship to the datum reference frame A, D, E. The meaning of the datums D and E (at MMC) relationship would then be mistaken in its interpretation unless qualified by analysis to invoke MMC. Further, a broad discrepancy could exist in such cases between a functional gage check versus a CMM check, for example.

DATUM FEATURES AT MMC, ZERO PERPENDICULARITY TOLERANCE

When an ensured closer relationship is desired between related features and their "size" datum features, several methods may be used. Applying "zero perpendicularity" at MMC as an orientation refinement of position tolerance, as shown in Fig. (b), is one method. This would relate to datum D at MMC size, ⌀ .250 (MMC size and virtual condition identical) and ensure the closer relationship to datum D as the secondary datum and datum E as the tertiary datum. However, as noted in a representative functional gage for this requirement (Fig. (a), page 269), the datum E pin would need to also compensate for the possible ⌀ .003 *positional* tolerances by use of a diamond pin of ⌀ .250 MMC size by diamond (or flat) width to a value less than .244 in the D-E direction as shown.

See further discussion and illustrations on the following pages.

AS DRAWN

(a)

AS DRAWN

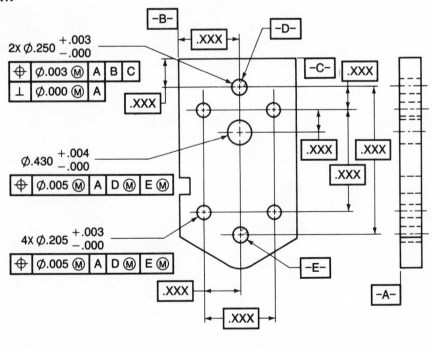

(b)

DATUM FEATURES AT MMC, ZERO POSITION TOLERANCE

(For discussion of Fig. (a) at right, see pages 266 and 267.)

The use of zero position tolerance further extends the principles as discussed in the foregoing paragraph. This is shown in Fig. (b) at right. The closer relationship of the five holes (ϕ .430 and ϕ .205) are now established with the refinement of both orientation and location. Note in the representative functional gage in Fig. (a) (page 271) that both of the pickup pins of the datums D and E holes are at *MMC size* (ϕ .250). Thus, the orientation question (as a part of position) requiring virtual condition recognition (Datum Virtual Condition Rule) is resolved. The MMC size and virtual condition are identical, ϕ .250 . The *closer* relationship is achieved.

Obviously, the use of zero position tolerance of any kind must be accompanied by good judgment, proper selection of controls, and due consideration of the need and impact on design and production. Zero position tolerancing is also discussed in other sections of this text.

See added discussion and illustrations on following pages.

GAGE

(a)

AS DRAWN

(b)

As discussed on the preceding page, the representative functional gage shown at right (Fig. (a)) would ensure the closer relationship of the five holes to the datums D and E holes. The datums D and E pickup pins are at ϕ .250 (MMC size) and, therefore, provide the closer relationship of the five hole pattern to the datums D and E (the datum reference frame A, D, E). That is, instead of ϕ .247 pins (D and E) as would be required in the earlier examples (using ϕ .003 pos tol at MMC), the movement of the pattern is more restricted. It is again evident that to ensure that the closer relationship results (pattern to datums), the orientation and position tolerance permitted on the datum features must be refined or restricted in some manner.

DATUM FEATURES AT MMC, COMPOSITE ZERO POSITION TOLERANCE

Figure (b) at right utilizes the composite tolerancing method as another option. This would provide the same results as discussed before except, of course, the more lenient location of the datum features D and E from the datum reference frame A, B, C. The reasoning and application of composite positional tolerance is covered in other portions of this text.

The functional gage shown above (Fig. (a)) would also verify this requirement as the feature relating tolerance for the five holes relative to datums D and E are the same as the preceding page example (Fig. (b)).

The upper entry, pattern locating tolerance (ϕ .030 at MMC) could be verified with a functional gage similar to that shown earlier in this section where A, B, and C surfaces (simulated datum planes) locate the part. The two gage pin sizes would be ϕ .220 (i.e., ϕ .250 MMC minus ϕ .030) to verify the locating holes (D and E).

See following pages for further possible applications of the two datum hole type of part.

GAGE

(a)

AS DRAWN

(b)

DATUM FEATURES CONTROLLED AT MMC, FEATURES RELATED TO DATUMS AT RFS

The illustration at right (Fig. (a)) controls the two ϕ .250 locations (datums D and E) on an MMC basis relative to the datum reference frame A, B, C. A functional gage identical to that shown in the first examples shown in this section could be used. Or, of course, comparable coordinate measuring machine method (CMM) with MMC analysis could be used. This example, as all others in this section, represents a mating part interface control with a mating part (with location pins).

As an addition to the possibilities discussed earlier which could be applied on parts of this kind, this example utilizes RFS on the datum features D and E for the five hole pattern relationship. The closer relationship of the five holes to the datum reference frame A, D, E is then achieved. That is, the effect of departure from MMC size of the datum features (D and E) and the commensurate movement of the five hole pattern relative to that departure is *not* permitted. A shift of the pattern (the five holes) could be detrimental to the two part interface at the datums (D and E) and, thus, the design restricts such movement by specifying RFS to the datums D and E. This will require the pattern to "center-up" on the secondary datum feature D and rotate ("square-up") on the tertiary datum feature E.

Figure (b) at right shows a representative functional gage to verify the A, D, E requirement. Note that a tapered or expanding pin is used to pick up the secondary datum feature hole ("fills" the hole) so that the part cannot shift. The Datum Virtual Condition Rule does not apply here because RFS has been specified on the datum feature D relationship. Rotation of the part is restricted by the RFS (tapered or expanding) pin to the tertiary datum feature E. Due to the potential positional error (ϕ .003 at MMC) of the D and E holes, some "give" must be included in Fig. (b). A diamond pin method could also have been used on the datum E pin if considered to be more practical than the sliding insert.

If a coordinate measuring machine or comparable inspection were used, part stabilization to the primary datum A, set-up on the secondary datum D feature RFS, and rotation against the datum E feature (RFS) would be accomplished identical to the principles of the gage shown at right. Although the RFS method on the secondary and tertiary datums is restrictive and should be carefully evaluated relative to the design, it does have the advantage of ensuring closer compatibility between the alternative methods of open set-up inspection.

See following pages for further discussion on two feature datum parts and other possibilities.

AS DRAWN

(a)

GAGE

(b)

DATUM FEATURES CONTROLLED AT RFS, FEATURES RELATED TO DATUMS AT MMC

In Fig. (a) at right another method is used to control the hole pattern ($\emptyset$ .430 and 4$\emptyset$ .205 holes) and the two hole datum (D and E) relationship.

Accuracy of the datum features D and E locations affects the relationship of the pattern of features related to them. In this case RFS has been applied to the $\emptyset$ .250 datums D and E holes to ensure greater precision of the datum features themselves. This reduces the error potential of the datum features relative to themselves (i.e., E to D), to the relationship of the five hole pattern, and also to the resulting relationship of *all* the holes to the outside surfaces of the part (datums B and C). That is, features D and E movement relative to datums B and C is restricted to $\emptyset$ .003, RFS. Thus, the five hole pattern in "following" datums D and E is also more restricted in its ultimate relationship to the outside surfaces.

Figure (b) at right illustrates that the verification of the $\emptyset$ .003, RFS requirement would be achieved in some form of open set-up measurement (i.e., coordinate measuring machine, etc.).

Since MMC relationships to datums D and E are specified on the five hole pattern relationship, a gage (and principles) as previously illustrated in this section could be used to verify that requirement. A virtual condition gage pin (to datums D and E) would be used.

Where yet further precision in the relationship is required, the zero perpendicularity or zero position tolerance methods (on an MMC basis, of course) could be specified. Further use of RFS principles could also be stated if necessary. See below for some added possibilities. Gages and comparable principles, as illustrated earlier in this section, would then be added to the considerations as necessary.

SOME ADDITIONAL POSSIBILITIES

Numerous other possibilities of the two hole datum method exist. Some are shown below. From the coverage of this section and of elsewhere in the text, these possibilities can be derived as best represents the design requirements.

AS DRAWN

(a)

(b)

In Fig. (a) shown at right a part (End Plate) is shown which uses a number of specifications and methods explained in earlier portions of this section and elsewhere in the text. Two of the four mounting holes (ϕ .200) are also utilized as the locating holes in the ultimate assembly. These holes will serve both to position and attach the part via the mating part holes and shoulder screws. The two selected datum holes are designated with the darkened quadrants ($\oplus$) and are specified as datums D and E. The four ϕ .200 holes are specified with a "pattern locating tolerance" from the datum reference frame A, B, C within ϕ .020, at MMC, using the composite tolerancing method. The primary datum plane A is established by datum target areas (A1, A2, A3) on the flat cast surface. This is to establish a plane of orientation which is functionally representative of the part interface (stabilize the related holes) to the mating part features, to minimize effect of the cast surface irregularities and warpage, as well as to clear the six bearing hole (ϕ .220, 4Xϕ .140 and ϕ .380) support lips. The secondary (B) and tertiary (C) datums are established from the indicated outside part edges. This positions the pattern appropriately on the part.

The two ϕ .200 holes ($\oplus$) used as the datums (D and E) are specified with zero (ϕ .000) tolerance, at MMC, as a "feature relating tolerance" (hole-to-hole in the pattern). This is to ensure adequate precision (orientation and location) of the locating datum holes and to enable the related six hole pattern to "follow" them with some precision. Earlier discussion in this section noted the reasons for such methods. The other two mounting holes (ϕ .200) are permitted a ϕ .010 positional tolerance, at MMC. This provides sufficient accuracy for these holes at assembly as clearance holes for the attaching screws.

The related six hole pattern features are given positional tolerances of ϕ .006, at MMC, relative to datums D and E, at MMC. Via the use of zero (ϕ .000) positional tolerancing relative to the datum features, the production and verification of these six holes will be to this precision as required by the design. For example, if a functional gage is used (see Fig. (b)) at right to simulate and verify this requirement, relationship ("pickup") with the datums D and E holes would be by MMC (same as virtual condition) size pins; ϕ .200 nominally. This situation has been discussed earlier in this section. Additionally, gage-maker's tolerances would be required as a further consideration.

If the part were verified by a coordinate measuring machine (CMM) or comparable methods, RFS pick up of datums D and E would probably be done first. Any permitted *pattern* movement relative to the datum feature's size departure from MMC would be determined, if necessary, by the amount of such departure. This would assume that the datum D hole orientation error (out-of-perpendicularity) did not negate the "bonus tolerance" and thus the permitted pattern movement. This would also assume that the datum E hole *size* was at least equal to, or greater than, the datum D hole size, and that it likewise did not negate the "bonus tolerance" and pattern movement via orientation (perpendicularity) *and* position error. The tertiary datum feature must not restrict the pattern relationship to the secondary datum feature. It must, however, provide the rotational orientation to stabilize the true position center planes (X, Y, and Z) and establish the basic pattern true positions. Graphic analysis, or computer analysis, can be used in lieu of gages. However, sizes of the holes and the amount of any orientation and/or positional error (quantified) of the datum features may have to be determined to complete such an evaluation. Since a functional gage (Fig. (b)) performs all necessary "judgments" on a "go," "no go" basis (is not quantified), no such exercise is then required. The trade-off negatives of the functional gage approach, however, must obviously also be considered.

AS DRAWN

NAME: END PLATE

(a)

GAGE

(b)

This form of specification to two hole datum features *does* achieve the "closer" pattern relationship discussed as a theme of this section. Of the numerous distinct options, the method which is most suitable is decided by the design requirements and any further influencing factors, i.e., production quantities, tooling and inspection costs, quality, reliability, time, etc.

Other gaging, or comparable verification (i.e., CMM), of the four ⌀ .200 hole pattern would be as shown in similar situations discussed in this section. The various methods shown earlier could have been applied to this part. The key determination is the desired design requirement, which will guide selection of the most appropriate method of control and the magnitude of the tolerances.

THE FUNCTIONAL GAGE

Earlier in this section functional gage principles were discussed generally and used to explain positional tolerancing options when relating to two size features (holes) as secondary and tertiary datums. Figure (b), page 277, illustrates a gage for the piece part shown in Fig. (a), page 277. Nominal sizes (i.e., virtual condition) gage pin sizes are given as developed from the part.

To build a functional gage, tolerance for the gage features location must be taken from the piece part feature location tolerance. This is commonly referred to as the 10% rule (or 5% to 10% rule), which means that up to 10% (sometimes slightly more) of the part tolerance limits could be used for gage tolerance. Therefore, a part of borderline (extreme limit of acceptable tolerance) conditions could be rejected by a functional gage if the part were at the fringe edge of the acceptable tolerance range. The gage will not, however, ever accept a "bad" part.

A functional gage has the following advantages:

1.   Minimizes time and resources involved to verify parts.

2.   Represents functional interface of the concerned features.

3.   Recognizes the subtle composite effects of size, orientation, and position as a "go," "no go" result.

4.   Provides a "hard" tool which can be utilized by anyone with reasonable technical skill; does not require a highly skilled inspector.

5.   Provides alternative methods for verification from surface plate, open set-up, coordinate measuring, etc.

6.   Will never accept a "bad" part.

Some disadvantages of a functional gage include:

1.   Requires gage-maker's tolerance *taken from* piece part tolerances (up to 10% usually).

2.   Could reject borderline good parts.

3.   Must be reworked if the part is revised.

4.   Costs for building, storage, and maintenance.

5.   Does not quantify results (it's "go" or "no go").

Application of gage design and make tolerances on the gage illustrated in Fig. (b), using the 10% rule, could vary depending upon the gage designer and any local practices. However, the following approach could be considered as one option:

Translating to coordinate dimensioning and tolerancing for gage design

Size: Datum pins and all gage pins +.0001 wear allowance and +.0001 size tolerance.

Location: Datum pin "E" and all gage pins ±.0001 from datum "D" pin in X and Y direction; consider diamond pin shape on datum "E" pin.

Location: Datum target pins A1, A2, A3 ±.003 from datum "D" pin in X and Y direction.

Basis for translation to gage pin tolerance within 10%:

ϕ .006 position tolerance on holes

10% of ϕ .006 = ϕ .0006 (Translated to ± coord tol = .00042)   **[SEE NOTES, BELOW]**

.0001 (WEAR) + .0001 (SIZE TOL) + .0002 (LOCATION TOL) = .0004

Slightly less than 10% of position tolerance ϕ .006 used

The preceding translation to X and Y dimensions, as could be applied to the illustrated gage, depicts the requirements in the language of the toolmaker. That is, the positional requirements of the piece-part, using geometrical tolerances, datums, etc., have been presented as directly relating to the gage-making. The number of gages being considered for construction is normally one, two, or very few. Geometrical tolerance methods, however, can also be used if considered desirable. Any advantages of doing so should be weighed. Someone, somewhere (probably the toolmaker) would be required to translate the "true position" values to coordinates to either make or verify the gage itself. Is this responsibility best placed with the designer, with the toolmaker, or with the gage inspector?

NOTE 1: It may be necessary to avoid this foreshortening of the position tolerance (to equivalent coordinate tolerance, i.e., 70% of ϕ .0006 = .00042) to retain sufficient tolerance to make the gage. For example, if .0006 is retained, the location of the gage pins (holes) could be ±.0002; (.0001 + .0001 + .0004 = .0006). This will, however, permit the possibility of slightly more than 10% of piece-part hole position tolerance (ϕ .006) to accumulate if the produced features of the gage deviate in location from the exact X and Y axes.

NOTE 2: This method uses the fail-safe distribution of position tolerance, permitting no more than .0002 coordinate error in X or Y about any one exact location. This is because the coordinate system and true position system do not equate exactly to each other (i.e., on which end of the coordinate tolerance of ±.0001 is the .0002 error, or is it equally distributed between, with .0001 on each end?) If the calculated risk were to be taken, hoping for or striving for equal distribution of the coordinate tolerance on *each* end, the gage-making pin (hole) location could specify ±.0002. Using baseline dimensioning as shown (all dimensions from one pin) could encourage such considerations with reduced risk.

If the geometrical tolerance method were used, the following could be considered for the illustrated gage:

Size: Datum pins and all gage pins +.0001 wear allowance +.0001 size tolerance.

Location: Datum pin "E" positional tolerance ϕ .0002, RFS, relative to datum "D" pin, RFS (and appropriate datum reference frame). All locating dimensions "basic."

Location: All gage pins positional tolerance ϕ .0004, RFS, relative to datums "D" and "E" pins, RFS (and appropriate datum reference frame). All locating dimensions "basic."

Location: Datum target pins A1, A2, A3 positional tolerance ϕ .004, RFS, relative to datum "D" and "E" pins, RFS (and appropriate datum reference frame).

All locating dimensions "basic."

Basis for translation to gage pin tolerance within 10%:

ϕ .006 position tolerance on holes

10% of ϕ .006 = ϕ .0006

.0001 (WEAR) + .0001 (SIZE TOL) + ϕ .0004 (LOCATION TOL) = ϕ .0006
10% of ϕ .006 used

The 10% rule and the feasibility of functional gaging is often governed by the tolerances which can be derived for gage-making. For example, the illustrated gage and its tolerances are ranging toward that fine limit which may become prohibitive in gage-making.* Therefore, the use of physical functional gages may not always be possible. However, optical or electronic techniques may be possible to achieve essentially the same results using comparators, special equipment, coordinate measuring machines, etc. The use of graphic analysis (paper gaging) from coordinate measured results provides a valuable option. See following sections of the text for explanation of such techniques.

Coordinate measuring machine data can be used to derive *mathematical* results which simulate functional gaging principles with great precision. In such cases no "gage tolerance" is used, although inherent machine error is present and may have to be considered in very precise applications.

Mathematical simulation of functional gaging principles to part holes and pins can be achieved via the use of calculators or computer programs. Of course, the knowledge of the principles involved and the computer programs must be present and available to perform such operations. Such techniques can be used with MMC principles involved (i.e., functional gaging principles) or with RFS principles. Results with these methods are derived rapidly, to great precision, at minimum cost, with minimal human error, and are able to be recorded in printouts, graphs, etc., with convenience and efficiency. Refer to section on Graphic Analysis and Mathematical Analysis of Coordinately Measured Features (page 281, etc.) in this text for further discussion of such techniques.

---

*See NOTES 1 and 2, page 279.

INTERRELATED DATUM REFERENCE FRAMES WITH COMMON DATUMS

In some cases a relationship between interrelated datums is established only to the extent of the common datum references. Hence separate datum reference frames may be established on the same part which are related only to the extent of their common datum references, with no further orientation implied or necessary; see the part shown at right.

MEANING

1. Several *separate* datum reference frames are established using common datums.

2. Datum C (slot) is a common tertiary datum for position toleranced hole patterns on each end of the part.

3. Datum D (hole) is a primary datum for runout of the step diameter, and is a secondary datum for runout of the counterbore and position of the slot.

AS DRAWN

GRAPHIC ANALYSIS AND MATHEMATICAL ANALYSIS OF COORDINATELY MEASURED FEATURES

Graphic analysis or mathematical analysis techniques may be applied as a preferred method or as an option to functional gaging where MMC is applied; or, also, on RFS applications, or coordinately measured part holes or pins.

Other sections of this text have addressed the principles of functional gaging and graphic analysis (paper gaging), e.g., pages 122–125, 130–135. This section expands the concept of graphic analysis and furthers the emphasis placed on mathematical methods as a viable option in part verification.

Suppose the part illustrated at right (Fig. (a)) has been produced and has been measured with a coordinate measuring machine (CMM) oriented and located to its datums A, B, C as shown below. The upper right illustration (Fig. (b)) shows a format which could be used to record the pertinent data obtained from the measurement and the graph (Fig. (c)) for "paper gaging." The circles (tolerance zones) could be a part of the graph or a transparent overlay.

The middle illustration (Fig. (d) at right) shows the four hole locational data recorded, and the dots on the graph are the plotted hole locations with respect to "true position" (at 0 in the center of the circular tolerance zones). The actual hole sizes are also recorded and the available position tolerance (with bonus tolerance), as based upon departure from MMC. From the graphic results, we can see that all holes are acceptable within their respective tolerance zones (using the bonus tolerance on holes 2, 3, 4). Obviously, accuracy of the plot is important, and the scale used is a factor in the precision. Human error plays a role, and some time and care is involved in the plot.

Closer scrutiny of the results, however, reveals holes 2 and 4 are very near their respective "no go" limits at their produced size. There could be a question as to whether the concerned hole is actually good; particularly if the plotted dot is so close to, or on, the limit circle of the tolerance zone. Mathematical verification of the hole's actual location (based upon a diametrical value) can be determined by use of a calculator or computer programs, as based upon the equation $Z = 2\sqrt{X^2 + Y^2}$ (as shown on page 304). The lower results, (Fig. (e)) as calculated, confirm that all holes *are* good mathematically, although holes 2 and 4 need to have been calculated only as necessary.

If the part had been functionally gaged but the gage did not "go," the graphic analysis or mathematical method could provide alternative methods of proof of verification. Or where a functional gage did not "go" and the graphic analysis could not clearly prove acceptability (dots near or on circular lines), the mathematical method could determine a final result. The mathematical method can, in fact, bypass both use of a functional gage *and* graphic analysis with great precision. Obviously, such results are based upon the precision of the coordinate measuring operation. Many CMMs are equipped with standard computer programs to perform such translations. Compact or portable computers or programmed calculators are also readily available to make such computations separately from the CMM computer. This provides portable convenience and less "tie-up" of the larger machines.

Graphic or mathematical analysis is advantageous for many reasons. Such methods provide preferred or optional methods to gaging, provide records which can "see" the results, provide data that can be used in tool maintenance or repair, requires no gaging tolerance, provides precision results, adapts to quality or inspection sampling procedures. The mathematical method provides the highest precision in evaluating the data, and in the shortest possible time.

4X Ø.300 +.005 −.000

⊕ | Ø.010 Ⓜ | A | B | C

(a)

HOLE NO., ACTUAL LOC., SIZE, POS TOL

① X =   Y =   SIZE-
                  POS TOL

② X =   Y =   SIZE-
                  POS TOL

③ X =   Y =   SIZE-
                  POS TOL

④ X =   Y =   SIZE-
                  POS TOL

(b)

(c)

4X Ø.300 +.005 −.000

⊕ | Ø.010 Ⓜ | A | B | C

HOLE NO., ACTUAL LOC., SIZE, POS. TOL

① X = .7465   Y = 2.249   SIZE − Ø.3025
     .7500   2.250   POS TOL Ø.0125
   (−).0035   (−).001

② X = .7456   Y = .746   SIZE − Ø.302
     .7500   .750   POS TOL Ø.012
   (−).0044   (−).004

③ X = 2.7522 Y = 2.2435   SIZE − Ø.304
     2.7500   2.2500   POS TOL Ø.014
   (+).0022   (−).0065

④ X = 2.755   Y = .754   SIZE − Ø.303
     2.750   .750   POS TOL Ø.013
   (+).005   (+).004

(d)

CALCULATED RESULTS

HOLE 1   Ø.00728011 Less than Ø.010 Tol zone

HOLE 2   Ø.01189285 Less than Ø.012 Tol zone (using bonus tol Ø.002)

HOLE 3   Ø.01372443 Less than Ø.014 Tol zone (using bonus tol Ø.004)

HOLE 4   Ø.01280625 Less than Ø.013 Tol zone (using bonus tol Ø.003)

(e)

Graphic analysis (paper gaging) of the part shown on pages 124 through 125 is repeated at right. Explanation of functional gaging and graphic analysis is also found on the referenced pages and on the preceding pages in this section.

Following are mathematical methods which can be used to determine the part acceptance, or not, in lieu of functional gages or graphic analysis:

STEP 1. Verify Pattern Locating Tolerance (ϕ .030) on three of four ϕ .250 holes. Using a programmed calculator or computer program*, each hole is *proven acceptable* as the following calculated results:

Hole 1 ϕ .03280244—Less than ϕ .034 Tol zone (using bonus tol ϕ .004)
Hole 2 ϕ .02683282—Less than ϕ .030 Tol zone
Hole 3 ϕ .02332381—Less than ϕ .030 Tol zone

STEP 2. Verify Feature Relating (hole-to-hole) Tolerance (ϕ .005) on four ϕ .250 holes. Using a computer program* developed with the required mathematical equations, the smallest diameter which will encompass all four produced hole centers while considering size departure from MMC as necessary is determined. The part (its four holes) is found to be within their respective tolerance zones as proven mathematically.

Using the data given at right on holes 1, 2, 3 and 4, and adding the bonus tolerance of ϕ .0015 on only hole 4 as shown, all holes are *proven acceptable*. The smallest diameter calculated, which will encompass all four hole centers based on the above conditions and as shown at right, is ϕ .0047472 (less than the ϕ .005 pos Tol). This value has been mathematically "corrected" for the MMC advantage of hole 4. This permits direct comparison to the stated ϕ .005 position tolerance requirement shown on the drawing. The corrected value, ϕ .0047472, is *less than* ϕ .005; therefore, all four holes are acceptable.

Coordinate measuring principles assumed as basis for data:

Step 1. Same as described in preceding example (pages 282, 283).

Step 2. As shown in illustration below.

*The computer programs are based upon the mathematics that is necessary to perform mathematical simulation of the hole location requirements. This simulation is also representative of a functional gage or graphic analysis (paper gaging). The Step 1 mathematical analysis can be performed by a calculator (See Appendix) or computer with the standard equation referenced. The Step 2 mathematical analysis requires a specially developed computer program. This must be based upon knowledge about geometric tolerancing principles, mathematics, and engineering judgment. Such capability now exists in The Lowell W. Foster Geometric Dimensioning and Tolerancing Module, "Geo-Calc," as used with a compact computer (T1 CC40); Step 1 uses Program 1; Step 2 uses Program 2. See "Acknowledgments" section in the front of the book.

STEP 1

OVERLAY GAGE (TOL ZONE)
.034
.032
.030
#2
#1
GAGE CENTER (TRUE POSITION)
DATUM B
#3
0
GRAPH
SCALE ■ = .001
DATUM C

1.000
3 4
1.500
.410
1 2
-C-
.525
4X ⌀.250 +.004 / -.000

⊕	⌀.030 Ⓜ	A	B	C
	⌀.005 Ⓜ	A		

HOLE

	X	Y	HOLE SIZE	TOL
①	.538 (Actual)	.420	⌀.254 (Actual)	⌀.030
	.525 (Drawing)	.410	.250 (MMC)	+.004
	.013 (+X)	.010 (+Y)	⌀.004	⌀.034
②	1.525	.422		⌀.030
	1.519	.410		
	.006 (−X)	.012 (+Y)		
③	.525	1.910		⌀.030
	.515	1.904		
	.010 (−X)	.006 (−Y)		

STEP 2

.005
.0055
.006
.0065
OVERLAY GAGE (TOL ZONE)
#1
SCALE ▨ = .001   ■ = .0002
GRAPH
0
#4
#2
#3

(OVERLAY GAGE)
.005
.0055
.006
.0065

1.000
3 4
1.500
.410
1 2
-C-
.525
4X ⌀.250 +.004 / -.000

⊕	⌀.030 Ⓜ	A	B	C
	⌀.005 Ⓜ	A		

	X	Y	HOLE SIZE	TOL
①	X = O (ORIGIN)	Y = O (ORIGIN)	GO/NO GO	⌀.005
②	1.0000 (−)	.0042	GO/NO GO	⌀.005
	.9988	.0000		
	.0012 (−X)	.0042 (−Y)		
③	X = O (SQUARE-UP)	1.5000	GO/NO GO	⌀.005
		1.4960		
		.0040 (−Y)		
④	1.0032	1.5000	⌀.2515 (Actual)	⌀.005
	1.0000	1.4965	⌀.2500 (MMC)	+.0015
	.0032 (+X)	.0035 (−Y)	⌀.0015 — —	⌀.0065

Graphic analysis (paper gaging) of the four hole pattern of the part shown on pages 129 and 131, and repeated at upper right, is shown in the lower right illustration. Explanation of the functional gaging and graphic analysis principles is found on pages 130 through 131 and preceding in this section. In this case, however, a datum feature of size (ϕ .500, datum "D") is referenced. In such a case, the four ϕ .380 holes must be located with respect to datum "D" and thus "follow" its location. Datum "A" and "B" surfaces are specified to give the hole pattern orientation and rotation to the mass of the part. The previously discussed principles of graphic analysis remain the same. However, the resulting center of the four hole pattern may "float" within a zone equal to the departure from MMC of datum "D".

GRAPHIC ANALYSIS

From the sample set-up and resulting data shown below, developed from a coordinate measuring operation, the graphic analysis of the four hole pattern (hole-to-hole) and the pattern relationship (center) is illustrated. This represents one of many possible solutions where the center of the pattern could float within the ϕ .0015 zone (per the datum hole ϕ .5015) and the hole centers also fall within their respective tolerance zones. As seen in the example, both hole-to-hole and pattern relationship to the datum "D" are acceptable within their respective zones.

	Measurement results					Bonus		Pos	
	in X	diff.	in Y	diff.	Hole size	Tol	+	Tol	
Hole 1	1.0025	(−) .0025	.3771	(−) .0021	ϕ .3825	ϕ .0025	+ ϕ .005	= ϕ .0075	
Hole 2	.9967	(+) .0033	.3735	(−) .0015	ϕ .382	ϕ .002	+ ϕ .005	= ϕ .007	
Hole 3	.9979	(−) .0021	.3738	(−) .0012	ϕ .3825	ϕ .0025	+ ϕ .005	= ϕ .0075	
Hole 4	.9972	(−) .0028	.3757	(−) .0007	ϕ .382	ϕ .002	+ ϕ .005	= ϕ .007	
Datum "D"	0	———	0	———	ϕ .5015	ϕ .0015		———	

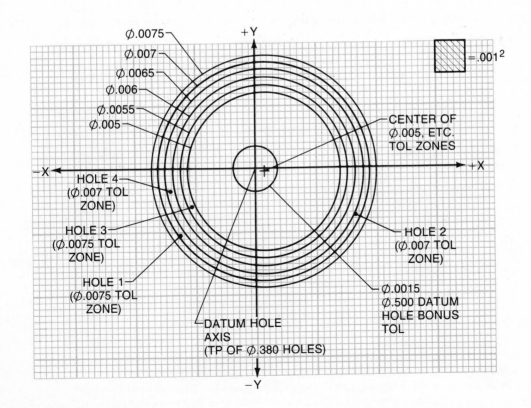

MATHEMATICAL ANALYSIS

Mathematical analysis of this part using the foregoing data (repeated below) can be utilized as a preferred method, or in lieu of graphic analysis (paper gaging) or even a functional gage.

Using a computer program* developed with the required mathematical equations, the smallest diameter which will encompass all four produced hole centers is determined. The program must also consider the effect of feature size departure from MMC of each of the four holes and also size departure from MMC of the datum feature and, thus, the permitted pattern float.

HOLE-TO-HOLE RELATIONSHIP

Based upon the data in the table on page 286 (including the bonus tolerances as shown), all holes *proved to be acceptable* in their *hole-to-hole* relationship and with respect to their tolerance zones about true position. The smallest diameter calculated which will encompass all four hole centers is ⌀ .0041522—less than ⌀ .005 stated on the drawing. This value has been mathematically "corrected" to take advantage of each hole departure from MMC. This permits a result which can be compared directly to the stated ⌀ .005 position tolerance requirement on the drawing.

HOLE PATTERN RELATIONSHIP

Based on the data in the table on page 286, the "hole pattern" also *proved to be acceptable* in its relationship to the datum "D" axis and its zone permitted by the datum feature "D"

*The computer programs are based upon the mathematics necessary to perform mathematical simulation of the hole location requirements. This simulation is also representative of a functional gage or graphic analysis (paper gaging). The programs must be based upon knowledge about geometric tolerancing principles, mathematics, and engineering judgment. Such capability now exists in The Lowell W. Foster Geometric Dimensioning and Tolerancing Module, "Geo-Calc," as used with a compact computer (TI CC40). See "Acknowledgments" section in the front of the book.

size departure from MMC. The result of the hole pattern location relationship calculation indicates the minimum bonus tolerance zone diameter required is ϕ .0002498—less than ϕ .0015 bonus tolerance provided by the datum ''D'' departure from MMC. See also page 290.

The graphic analysis method of page 287, and repeated below, ''pictures'' what has mathematically been done.

DATUM VIRTUAL CONDITION EFFECT

Additional "float" is available an amount equal to the difference between the ⌀ .500 MMC size and its "mating size"* (the largest inscribed cylinder, perpendicular to datum A which can be brought into contact against the ⌀ .500 datum A feature). This recognizes the possible virtual condition size of ⌀ .497 as invoked by the Datum Virtual Condition Rule and as found on the functional gage shown on page 131. For example, if the "mating size"* was ⌀ .498, an *additional "float"* of the center of the four hole pattern could be ⌀ .002 (⌀ .500 MMC − ⌀ .498 = ⌀ .002). Thus, the total float ("Bonus tolerance") of the example part described in the data above and illustrated on the graphic analysis on Pages 287 and 289 could be up to ⌀ .0035; ⌀ .0015 (departure from MMC size) + ⌀ .002 (difference between ⌀ .500 MMC size and ⌀ .498 "mating size"*); i.e. ⌀ .0015 + ⌀ .002 = ⌀ .0035. The ⌀ .0035 value could have been used in the illustrated graphic analysis and in the mathematical analysis (Page 288) as the "bonus tolerance" (instead of ⌀ .0015 as shown) on this part as necessary. However, in this example of possible results in a hypothetical situation, the part was found acceptable without this consideration and the added inspection steps (determine "mating size"*) were not needed.

---

*"Mating size", a term used in ISO (International Standards Organization) standards which is not found in ANSI Y14.5M—1982, is used here to describe this situation.

CONCENTRICITY

CONCENTRICITY ⊚

Definition. Concentricity is the condition where the axes of all cross-sectional elements of a feature's surface of revolution are common to the axis of a datum feature.

Concentricity tolerance. Concentricity tolerance is the diameter of the cylindrical tolerance zone within which the axis of the feature(s) must lie; the axis of the tolerance zone must coincide with the axis of the datum feature(s).

AS DRAWN

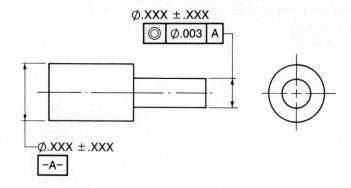

Ø.XXX ± .XXX

⊚ | Ø.003 | A

Ø.XXX ± .XXX

-A-

MEANING

⊚ | Ø.003 | A

TO DATUM A
REGARDLESS OF
DATUM SIZE

WITHIN Ø.003 RFS

THIS FEATURE SHALL
BE CONCENTRIC

DATUM DIA.
A
(RFS)

DATUM
AXIS

Ø.003
TOL ZONE
(RFS)

Ø.003
TOL ZONE
(RFS)

DERIVED AXIS OF
FEATURE ELEMENTS
MUST BE WITHIN
Ø.003 RFS

Concentricity is a type of location tolerancing. It always involves two or more basically coaxial features of size and controls the amount by which the *axes* of the features may fail to coincide. Concentricity tolerance, due to its unique characteristic, is always used on an RFS basis.

Where interrelated features are basically coaxial, we must first consider the possibility of using the more economical position (MMC) or runout controls before considering concentricity. (See also "Coaxial Features—Selection of Proper Control" page 298.)

Concentricity relates to a condition in which two or more features (cylinders, cones, spheres, hexagons, etc.) in any combination have a common axis. Concentricity tolerance is the diameter of the cylindrical tolerance zone within which the axis of the feature (or features) must lie, the axis of the tolerance zone must coincide with the axis of the datum feature or features.

The surface of a feature must be used to establish its axis. Therefore, all the irregularities or errors of form of a feature surface must be considered in establishing the axis. For instance, the surface may be bowed, out of circularity, etc., in addition to being offset from its datum feature. This usually involves a complex inspection analysis of the entire surface and therefore requires a more time-consuming and costly procedure.

Concentricity requirements are required less frequently than position (MMC) or runout requirements. However, where concentricity is required, it provides effective control over the unique applications of coaxial relationships. For example, concentricity might be applied to the coaxiality requirements of a tape drive pulley or of a capstan on a computer mechanism or a motor generator rotor. Often where balance is required, the out-of-circularity or lobing effect (and possible other form errors) may be permissible although it may exceed the conventional FIM requirement. Hence, any basically symmetrical form of revolution (hexagons, cones, etc.), or consistently symmetrical variation of such shape, could satisfy a concentricity tolerance where a runout requirement may not.

As stated before, the axis of a feature is determined by the *surface* of that feature. We compare that feature axis with the datum feature axis; the result of this comparison will determine whether or not the feature axis satisfies the concentricity requirement.

A common method of determining the coaxiality of a feature with its datum (such as in runout) uses dial indicator (FIM) readings, with the part mounted and rotated on its datum feature. This method is also often adequate as a means of determining whether concentricity requirements have been satisfied. However, in concentricity, by the very nature of its characteristics, a FIM reading in excess of the stated permissible tolerance may not necessarily mean that the feature *axis* lies outside the concentricity tolerance zone. That is, if a feature has a surface error such as out-of-circularity, its surface will register a FIM dial reading relative to its datum diameter but its axis may not have actually exceeded the concentricity tolerance zone.

Figures 1 through 5 illustrate this comparison with FIM and the principles involved in concentricity tolerancing. Using the sample requirement shown, Figure 1 shows a part that is theoretically perfect; it also shows the location of the cylindrical concentricity tolerance zone about the datum axis.

Figure 2 illustrates the part feature at its maximum displacement (or combination of various form errors). Note that the displacement registered by .003 FIM at the *surface* is equivalent to the *diametral* displacement of the *axis* within the concentricity tolerance zone. Figure 3 illustrates an out-of-circularity condition for which the .003 FIM reading is also *within* the specified concentricity zone. Under the conditions illustrated in Figs. 2 and 3, since the FIM is within the stated concentricity tolerance, the parts are acceptable with no further considerations. In this case, the results are essentially identical to a requirement which might have been stated using total runout on the basis of the FIM only. Therefore, to some extent, concentricity and runout can be considered identical except for interpretation, that is, surface variation (FIM direct) vs. axial displacement (as derived from FIM or some other comparable method).

Also, the above provides the uniform interpretation of any drawings for which concentricity (RFS) as previously defined (now generally equivalent to a total runout requirement) is used. However, where the stated concentricity tolerance is exceeded by the FIM, the characteristics of concentricity as now interpreted may possibly authorize acceptance of the part as being concentric within the specified limits.

Figure 4 illustrates the part with a .005 FIM reading which appears to exceed the stated concentricity tolerance. Yet closer observation will show that the part is actually *perfectly concentric* (as a hypothetical example) because the out-of-circularity error (possibly including other form errors) is symmetrically uniform. The FIM reading as conventionally applied is no longer conclusive and, in fact, could reject good parts. Closer scrutiny of Fig. 3 also shows that the FIM reading is again inconclusive with respect to the part's actual concentricity. The part is perfectly concentric although a FIM reading was registered.

Therefore, whenever a part is being evaluated for concentricity with conventional FIM techniques and is found to be within the concentricity tolerance stated, the part has met the concentricity tolerance requirement. However, when the part *exceeds* the FIM as conventionally checked or when actual concentricity is to be checked, more extensive inspection techniques are required, i.e., further consideration must be given to the question of whether additional form errors, such as out-of-circularity, are permissible within the concentricity tolerance. The form errors may vary anywhere within the tolerances of size so long as the concentricity tolerance is met.

Figure 5 illustrates the principles involved in checking actual concentricity. A precision collet and indicator set-up method are illustrated for purposes of explanation. However, other methods can be used, such as precision spindle inspection machines which are capable of picking up the true axes of rotation.

In the method shown, a measurement (M1) is taken of the part feature at a given point (X). The part is rotated 180°, and another measurement (M2) is taken.

COMPARISON WITH FIM METHODS

MEANING

AS DRAWN

◎ | ∅.003 | A

0 FIM

∅.003 ◎
TOL
ZONE

DATUM
A
AXIS

DATUM A

FIGURE 1

.003 FIM

∅.003
TOL ZONE

FIGURE 2

.003 FIM

∅.003
TOL ZONE

FIGURE 3

.005 FIM

∅.003
TOL ZONE

FIGURE 4

M₂

KNOW DISTANCE K

M₁

PART MOUNTED
(PRECISION COLLET ETC.)

POINT
X

POINT
X

DATUM AXIS

$(M_2 - K) - (K - M_1) =$ DEVIATION FROM ◎
DIAL INDICATOR READINGS TAKEN ⊥ TO DATUM AXIS

FIGURE 5

The difference between these readings (see formula) is the deviation from concentricity (as a diameter) at that cross-sectional plane which is perpendicular to the datum axis.

Measurements are made at as many diametrical cross sections on the entire surface as necessary to adequately satisfy the requirement. For part acceptance, the maximum deviation from the concentricity calculated should be within the permissible concentricity tolerance. Since the deviation from the concentricity calculated is expressed as a diameter, it can be compared directly with the stated concentricity tolerance zone.

As previously described, when checking concentricity, we consider all the irregularities of errors of form of a feature surface in composite to determine its axis, and thus the displacement of this axis relative to the datum axis.

Since the dial indicator (or other measuring method) sweeps the surface, the type of errors involved are, for all practical purposes, indistinguishable from one another.

To illustrate the various basic types of error involved, the example shown on the next page explains the general effect of feature eccentricity, error of parallelism of axis, out-of-straightness of axis, out-of-circularity, and out-of-cylindricity on concentricity tolerance requirements using FIM methods. Note that the sample part has a concentricity tolerance requirement of ⌀ .001.

Eccentricity is seen as the allowable parallel displacement of the feature axis with respect to the datum axis. The .0005 eccentricity is a displacement to one side but must be considered as a ⌀ .001 total tolerance cylinder.

Out-of-straightness and error of parallelism of axis is seen as a tip up, down, sideways, or bow of the feature with respect to the datum axis. The ⌀ .001 cylindrical tolerance zone represents the limits of the allowable error of this type.

Out-of-circularity and out-of-cylindricity (as it affects concentricity) is the difference in the diameters of two concentric circles and cylinders (in planes normal to the axis) between which the actual surface of the feature must fall.

As is seen, concentricity tolerance is a composite control of all elements of the feature concerned as they establish an axis. It is this *axis* which serves as the design requirement and thus its basis for part evaluation or inspection.

For all practical purposes, when concentricity is the requirement, these errors are viewed as indistinguishable from one another; as a result they are considered in composite when we determine whether the part complies with the concentricity tolerance.

A study of the principles of concentricity and its comparison with other coaxial controls (i.e., runout or position) will reveal a duplicity. As is possible with some other geometric tolerances, alternate symbolic representation of concentricity could be described as a positional tolerance with duplicate meaning. The concentricity callout on the opposite page could be specified as shown below with the same meaning of:

AS DRAWN

MEANING

COAXIAL FEATURES—SELECTION OF PROPER CONTROL

There are three methods of controlling interrelated coaxial features:

1.   RUNOUT TOLERANCE (circular or total)   (RFS)

2.   POSITION TOLERANCE   (MMC)

3.   CONCENTRICITY TOLERANCE   (RFS)

Any of the above methods provide effective control. However it is important to select the *most appropriate* one to both meet the design requirements and provide the most economical manufacturing conditions. (See also details of preceding sections).

Below are recommendations to assist in selecting the proper control:

If the need is to control only CIRCULAR cross sectional elements in a composite relationship to the datum axis, RFS, e.g., multi-diameters on a shaft, use

CIRCULAR RUNOUT ✗   **EXAMPLE**

(This method controls any composite error effect of circularity and circular cross-sectional profile variations)

If the need is to control the TOTAL cylindrical or profile surface in composite relative to the datum axis' RFS, e.g., multidiameters on a shaft, bearing mounting diameters, etc., use

TOTAL RUNOUT ✗✗   **EXAMPLE**

(This method controls any composite error effect of circularity, cylindricity, straightness, angularity, and parallelism.)

NOTE   Runout is always implied as an RFS application. It cannot be applied on an MMC basis, since an MMC situation involves functional interchangeability or assemblability (probably of mating parts), in which case POSITION tolerance would be used. See below.

If the need is to control the total cylindrical or profile surface and its axis in composite location relative to the datum axis on an MMC basis, e.g., on mating parts to assure interchangeability or assemblability, use

POSITION ⊕   **EXAMPLE**

If the need is to control the *axis* of one or more features in composite relative to a *datum axis*, RFS, e.g., to control balance of a rotating part, use

CONCENTRICITY ◎   **EXAMPLE**   ◎ | ⌀.002 | A–B

NOTE   Concentricity is always implied as an RFS application. Variations in size (departure from MMC size, out-of-circularity, out-of-cylindricity, etc.) do not in themselves conclude *axis* error.

AS DRAWN

MEANING

The example above illustrates a concentricity tolerance, using two datums to establish a common datum axis.

The 1.000 ± .001 outside diameter is to be concentric to the common datum axis established by datum diameters A and B, within ⌀ .004 regardless of the sizes to which any of the diameters are produced. The diameter size tolerances must be evaluated separately.

The lower portion of the illustration shows the part mounted on both datum diameters A and B, so that the concentricity of the 1.000 diameter with respect to both diameters A and B and thus to their common axis can be checked in one operation. The FIM method is not conclusive when the reading exceeds the stated concentricity tolerance. In this case, other methods must be used (see previous pages) to determine whether the feature does exceed the concentricity tolerance in an actual concentricity check.

APPENDIX

APPENDIX

CONVERSION OF ⊕ POSITION (CYLINDRICAL) TOLERANCE ZONES TO/FROM COORDINATE TOLERANCE ZONES

⊕ TO ±

TOTAL ⊕ TOL ZONE x .70711 = TOTAL COORDINATE TOLERANCE ZONE

EXAMPLE: .007 ⊕ TOL x .70711 = .00495 { .005 TOTAL COORDINATE TOL
OR
.0025 BILATERAL TOL

RULE OF THUMB:
USE .7 (OR 70%) OF TOTAL ⊕ TOL ZONE TO CONVERT QUICKLY IN NON-CRITICAL APPLICATIONS, e.g. .7 x .007 = .0049 OR .005 (±.0025)

.007 TOTAL ⊕ POSITION TOLERANCE ZONE DIAMETER

TOTAL COORDINATE OR BILATERAL TOLERANCE ZONE

± TO ⊕

TOTAL COORDINATE TOL ZONE x 1.4142 = TOTAL ⊕ TOL ZONE

EXAMPLE: .005 TOTAL COORDINATE TOL
OR
.0025 BILATERAL TOL x 2
} x 1.4142 = .007 TOTAL ⊕ TOL

RULE OF THUMB:
USE 1.4 TIMES TOTAL COORD TOL ZONE TO CONVERT QUICKLY IN NON-CRITICAL APPLICATIONS, e.g. 1.4 x .005 = .007 ⊕ TOL

CONVERSION CHART

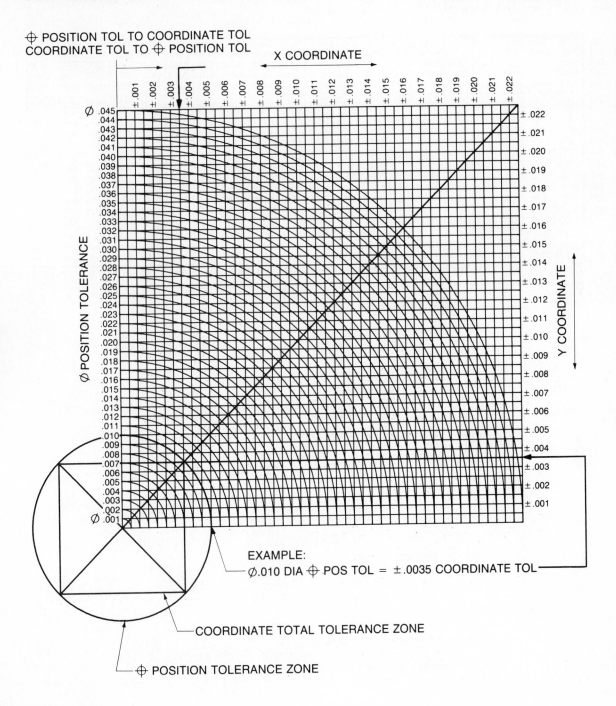

⊕ POSITION TOL TO COORDINATE TOL
COORDINATE TOL TO ⊕ POSITION TOL

X COORDINATE

∅ POSITION TOLERANCE

Y COORDINATE

EXAMPLE:
∅.010 DIA ⊕ POS TOL = ±.0035 COORDINATE TOL

COORDINATE TOTAL TOLERANCE ZONE

⊕ POSITION TOLERANCE ZONE

CONVERSION OF COORDINATE MEASUREMENTS TO ⊕ POSITION LOCATION

Y	.001	.002	.003	.004	.005	.006	.007	.008	.009	.010	.011	.012	.013	.014	.015	.016	.017	.018	.019	.020
.020	.0400	.0402	.0404	.0408	.0412	.0418	.0424	.0431	.0439	.0447	.0456	.0466	.0477	.0488	.0500	.0512	.0525	.0538	.0552	.0566
.019	.0380	.0382	.0385	.0388	.0393	.0398	.0405	.0412	.0420	.0429	.0439	.0449	.0460	.0472	.0484	.0497	.0510	.0523	.0537	.0552
.018	.0360	.0362	.0365	.0369	.0374	.0379	.0386	.0394	.0403	.0412	.0422	.0433	.0444	.0456	.0469	.0482	.0495	.0509	.0523	.0538
.017	.0340	.0342	.0345	.0349	.0354	.0360	.0368	.0376	.0385	.0394	.0405	.0416	.0428	.0440	.0453	.0467	.0481	.0495	.0510	.0525
.016	.0321	.0322	.0325	.0330	.0335	.0342	.0349	.0358	.0367	.0377	.0388	.0400	.0412	.0425	.0439	.0452	.0467	.0482	.0497	.0512
.015	.0301	.0303	.0306	.0310	.0316	.0323	.0331	.0340	.0350	.0360	.0372	.0384	.0397	.0410	.0424	.0439	.0453	.0469	.0484	.0500
.014	.0281	.0283	.0286	.0291	.0297	.0305	.0313	.0322	.0333	.0344	.0356	.0369	.0382	.0396	.0410	.0425	.0440	.0456	.0472	.0488
.013	.0261	.0263	.0267	.0272	.0278	.0286	.0295	.0305	.0316	.0328	.0340	.0354	.0368	.0382	.0397	.0412	.0428	.0444	.0460	.0477
.012	.0241	.0243	.0247	.0253	.0260	.0268	.0278	.0288	.0300	.0312	.0325	.0339	.0354	.0369	.0384	.0400	.0416	.0433	.0449	.0466
.011	.0221	.0224	.0228	.0234	.0242	.0250	.0261	.0272	.0284	.0297	.0311	.0325	.0340	.0356	.0372	.0388	.0405	.0422	.0439	.0456
.010	.0201	.0204	.0209	.0215	.0224	.0233	.0244	.0256	.0269	.0283	.0297	.0312	.0328	.0344	.0360	.0377	.0394	.0412	.0429	.0447
.009	.0181	.0184	.0190	.0197	.0206	.0216	.0228	.0241	.0254	.0269	.0284	.0300	.0316	.0333	.0350	.0367	.0385	.0402	.0420	.0439
.008	.0161	.0165	.0171	.0179	.0189	.0200	.0213	.0226	.0241	.0256	.0272	.0288	.0305	.0322	.0340	.0358	.0376	.0394	.0412	.0431
.007	.0141	.0146	.0152	.0161	.0172	.0184	.0198	.0213	.0228	.0244	.0261	.0278	.0295	.0313	.0331	.0349	.0368	.0386	.0405	.0424
.006	.0122	.0126	.0134	.0144	.0156	.0170	.0184	.0200	.0216	.0233	.0250	.0268	.0286	.0305	.0323	.0342	.0360	.0379	.0398	.0418
.005	.0102	.0108	.0117	.0128	.0141	.0156	.0172	.0189	.0206	.0224	.0242	.0260	.0278	.0297	.0316	.0335	.0354	.0374	.0393	.0412
.004	.0082	.0089	.0100	.0113	.0128	.0144	.0161	.0179	.0197	.0215	.0234	.0253	.0272	.0291	.0310	.0330	.0349	.0369	.0388	.0408
.003	.0063	.0072	.0085	.0100	.0117	.0134	.0152	.0171	.0190	.0209	.0228	.0247	.0267	.0286	.0306	.0325	.0345	.0365	.0385	.0404
.002	.0045	.0056	.0072	.0089	.0108	.0126	.0146	.0165	.0184	.0204	.0224	.0243	.0263	.0283	.0303	.0322	.0342	.0362	.0382	.0402
.001	.0028	.0045	.0063	.0082	.0102	.0122	.0141	.0161	.0181	.0201	.0221	.0241	.0261	.0281	.0301	.0321	.0340	.0360	.0380	.0400

X →

ACTUAL LOCATION

X — DIFFERENTIALS

Y

EQUATION:
$$Z = 2\sqrt{X^2 + Y^2}$$

TRUE POSITION

ACTUAL LOCATION

DIAMETRICAL EQUIVALENT

DATUM PLANE

90°

DATUM PLANE

CONVERSION OF COORDINATE MEASUREMENTS TO ⊕ POSITION LOCATION—CALCULATOR

.00943398

PROGRAM:
$$Z = 2\sqrt{X^2 + Y^2}$$

Z = DIAMETRICAL EQUIV. OF X & Y DIFFER. IN MEASUREMENTS

MAY BE DONE BY CALCULATOR OR COMPUTER

EXAMPLE

3x Ø.250 $^{+.005}_{-.000}$

$\oplus$ | Ø.010 Ⓜ | A | B | C

CONVERSION

PRODUCED HOLE .250 (MMC)
($\oplus$ ZONE = .010)
ACTUAL HOLE CENTER

ACTUAL MEASUREMENT

(HORIZONTAL)
ACTUAL − BASIC = X
.754 − .750 = .004

(VERTICAL)
BASIC − ACTUAL = Y
.600 − .598 = .002

FROM CHART ABOVE .004 (X)
AND .002 (Y) GIVE .0089 DIA.
HOLE LOCATION IS WITHIN
SPECIFIED .010 DIA. HOLE LOCATION
ACCEPTABLE

PRODUCED HOLE AT .255 (MIN MC)
($\oplus$ TOL ZONE = .015 (.010 + .005))
ACTUAL HOLE CENTER

ACTUAL MEASUREMENT

(HORIZONTAL)
ACTUAL − BASIC = X
.756 − .750 = .006

(VERTICAL)
BASIC − ACTUAL = Y
.600 − .596 = .004

FROM CHART ABOVE .006 (X)
AND .004 (Y) GIVE .0144 DIA.
HOLE LOCATION IS WITHIN
PERMISSIBLE .015 DIA WITH HOLE AT
HIGH LIMIT. HOLE LOCATION ACCEPTABLE.

USA PRACTICES AND INTERNATIONAL PRACTICES

With the advent of ANSI Y14.5M–1982, an improved base of compatability exists between USA and ISO (International) practices. Further, with the current U.S. involvement and input to revisions and expansion of international standards, even greater improvement will evolve in the future.

Where there is concern that U.S. practices be implemented for greatest possible compatibility with international practices, the following points of discussion may be helpful.

MMC AND RFS CONDITIONS

Referring to RULE 2, it is noted that the condition desired (MMC or RFS) must be specified on all applications of POSITION tolerance and appropriate related datums as preferred practice. This achieves a degree of agreement with international practices. ISO practices require MMC be specified when desired; no indication of condition implies regardless-of-feature-size application. Incorporating other new refinements and options of Y14.5M–1982, the below comparison and similarities are found.

USA (ANSI Y14.5M–1982)   **ISO 1101**

MMC

(or metric application as shown in Y14.5M–1982)

USA(ANSI Y14.5M–1982)   **ISO(ISO1101)**

RFS

(or metric application as
shown in ANSI Y14.5M–1982)

The Ⓢ condition symbol is required in U.S. practices as retained from previous standards to assure uniform interpretation. It is anticipated that future expansion of U.S. standards may drop the need for Ⓢ. Datum precedence and the vertical line separation in the symbol is international practice.

ISO 1101 PRACTICES

The decimal comma (,) is used in some metric countries as shown on the following page. This principle could be used to no prejudice of the illustration in this text. These examples also illustrate ISO 1101 datum letter sequence and datum identification triangle symbol (⊥) use.

FIGURE 1

FIGURE 2

FIGURE 3

COMPARISON BETWEEN ANSI Y14.5M–1982 AND ISO/1101–1983

CHARACTERISTIC	ANS1—Y14.5M	ISO—1101
STRAIGHTNESS	—	—
FLATNESS	▱	▱
ANGULARITY	∠	∠
PERPENDICULARITY (SQUARENESS)	⊥	⊥
PARALLELISM	//	//
CONCENTRICITY	◎	◎
POSITION	⊕	⊕
CIRCULARITY (ROUNDNESS)	○	○
SYMMETRY	USE ⊕	≡
PROFILE OF A LINE	⌒	⌒
PROFILE OF A SURFACE	⌓	⌓
RUNOUT (CIRCULAR)	↗ ↗	↗
RUNOUT (TOTAL)	↗↗ ↗↗	↗↗
CYLINDRICITY	⌭	⌭
DATUM FEATURE	-A-	A ◢
MAXIMUM MATERIAL CONDITION (MMC)	Ⓜ	Ⓜ
REGARDLESS OF FEATURE SIZE (RFS)	Ⓢ	NONE (ASSUMED UNLESS) SPECIFIED MMC
LEAST MATERIAL CONDITION (LMC)	Ⓛ	(PROPOSED)
DATUM TARGET	(A1)	(A1)
PROJECTED TOLERANCE ZONE	Ⓟ	Ⓟ

TANGENT PLANE*

AS DRAWN

TANGENT PLANE

MEANING

TANGENT PLANE—
PLANE OF CONTACT ON SURFACE
EXTREMITIES MUST BE WITHIN
.003 PERPENDICULARITY TOL ZONE

.003 TOL ZONE→

90°

DATUM PLANE A

Where an orientation tolerance (e.g. Perpendicularity) is critical *only* to those portions of a flat surface which may be in contact with a mating part as a plane of tangency, the *Tangent Plane* notation is added beneath the feature control frame. The resulting tangent plane (e.g. inspection plate or representation) only, must be within the orientation tolerance zone.

---

*NOTE   This is advisory only; *not* per ANSI Y14.5M—1982.

FREE–STATE VARIATION: "RESTRAINED" APPLICATION

Free-state variation is the amount a part distorts after removal of extrenal forces applied during manufacture, for instance, parts consisting essentially of shells or tubes with a thin wall thickness in proportion to the diameter. Geometric tolerances (such as circularity, cylindricity, and concentricity) cannot be properly applied without controlling free-state variation on parts of this type.

Variations in the free state can exist in two ways: (1) distortion due to the weight or flexibility of the part, or (2) distortion due to internal stresses set up in fabrication.

Where free-state variation control is necessary, any datums and the features in control may require specification of their allowable free-state variation or the maximum force necessary to restrain each of them to drawing tolerance—so that the desired assembly relationship can be stated and represented for evaluation of the requirement, thus achieving compliance.

EXAMPLE

⚠APPLIES IN FREE STATE WHEN DATUM ⌀ A IS RESTRAINED TO DRAWING TOLERANCE
AND SURFACE B IS HELD FLAT WITH THE FORCE OF 150/175 POUNDS WITH PART
IN HORIZONTAL POSITION.

MEANING

Compliance with the drawing note as stated.

AVERAGE DIAMETER

An average diameter is the mean of several diameters (not less than four) across a circular or spherical part used to determine conformance to *diameter* tolerance only. If practicable, the average diameter may be determined by using a periphery tape. Only when a diameter is allowed a maximum circularity tolerance in the free state should it be specified as AVG $\emptyset$.

A part of this kind is normally expected to flex to proper form or shape at assembly. The reason for control in its free state is to facilitate hand or automated assembly and handling while the part is in its free-state configuration, to restrict distortion to within safe elastic limits of the material, etc.

EXAMPLE

MEANING

1.   The high and low diameter size measurements (four minimum) must differ by no more than .060 (because tolerance is .030 round on R). The calculated *average* (mean) of these diameters must be within the specified size tolerance 13.490–13.510 diameters.

2.   The entire surface at the cross section measured must lie within the .030 wide tolerance zone.

If the part in free state is of the general shape shown in the figure below, the 13.540 could be the highest diameter measurement, and 13.480 could be the lowest diameter measurement.

For example, four measurements are:

13.540	(high)
13.525	
13.495	
13.480	(low)
54.040	

13.510 average diameter (within *size* tolerance)

$\emptyset$13.540 $\emptyset$13.480

.030

13.540 −13.480 =.060
AVERAGE DIA =13.510

High and low measurements do not differ by more than .060; therefore, part is within free-state *circularity* tolerance.

If the part in free state is of the general shape shown in the figure below, the 13.520 could be the highest diameter measurement, and 13.460 could be the lowest diameter measurement.

For example, four measurements taken are:

13.520 (high)
13.500
13.480
13.460 (low)
―――――
53.960

13.490 average diameter (within *size* tolerance)

13.520 − 13.460 = .060
AVERAGE DIA = 13.490

CLARIFICATION OF RULE 1 OF ANSI Y14.5

That Rule 1 applies to "individual" features *only* and *not* to an "interrelationship" of individual features and that form variation of the individual feature is contained within the PERFECT FORM AT MMC BOUNDARY will be explained in detail in the following text and illustrations.

Rule 1 applies to "individual" features whether the form variation is through permissible SIZE variation (form tolerance *not* specified) or whether the tolerance is specified.

The illustrations on the following pages show that the individual features involved in the form tolerances of FLATNESS, STRAIGHTNESS, CIRCULARITY, CYLINDRICITY and the orientation tolerance PARALLELISM readily comply with the definition of "individual" feature (see definition in section of "Clarification of Feature") and are to be within the perfect form at MMC boundary. If no form tolerance relationship had been specified, we would nonetheless assume that these features were also confined to the perfect form at MMC boundary.

ANGULARITY is an orientation characteristic involving an *interrelationship* of individual features. Rule 1 does *not* apply to such interrelationship, so angularity control must be specified. However, an extremity of the angular surface which is adjacent to, or part of, a surface which is size-dimensioned must be contained within the perfect form at MMC boundary, such as, for example, the lower left surface extremity near the angle vertex in the figure illustrating angularity.

PERPENDICULARITY is an orientation characteristic and defines an interrelationship of individual features; thus Rule 1 does *not* apply. Where perpendicularity is required, it must be specified. However, the individual surface features must be contained within the perfect form at MMC boundary as established by the size dimensions (such as the X and Y dimensions shown).

A PROFILE OF A SURFACE tolerance is usually applied as a combined form, orientation and size control (i.e., a BASIC overall dimension with a profile of surface tolerance). In this case, it is *not* subject to Rule 1. Where a PROFILE OF A LINE control is used as a refinement of size control, Rule 1 *would* apply.

RUNOUT, being a unique variety of rotating surface tolerance, involves an interrelationship of individual features; thus Rule 1 does *not* apply. However, the individual features must be confined to their size dimensions and perfect form at MMC boundary.

From the preceding we can state in summary that unless otherwise specified or controlled, the "size" limits of an individual feature control the applicable "form" variation within these size limits and the boundary of perfect form when no form tolerance is stated; furthermore, when a form tolerance *is* specified, the form variation must also be contained within these size limits and the boundary of perfect form at MMC (exception: straightness specified on ϕ basis and related to size dimension). Rule 1 does not control and is not intended to control "interrelationships" of individual features. When an "interrelationship" is required, it must be specified.

PERFECT FORM AT MMC AS APPLIED TO FORM, ORIENTATION, AND PROFILE TOLERANCE

PERFECT FORM AT MMC AS APPLIED TO FORM, ORIENTATION, AND RUNOUT TOLERANCE

PERF FORM AT MMC BOUNDARY
APPLIES TO
X & Y SEPARATELY (INDIVIDUAL FEATURES).

INTERRELATIONSHIP
CONTROLLED BY ORIENTATION
TOL AS STATED

PERFECT FORM AT MMC
ENV. APPLIES TO EACH
FEATURE SEPARATELY.

INTERRELATIONSHIP
CONTROLLED BY RUNOUT
TOL

CLARIFICATION OF FEATURE

The meaning of the term "feature" is defined below. The definitions clearly show the flexibility of this term.

Definitions.

Features. A feature is the general term applied to a physical portion of a part and may include one or more surfaces such as holes, pins, screw threads, profiles, faces, or slots. A feature may be "individual" or "related."

Individual Feature. The "individual" feature, as implied by Rule 1, is a feature related to the part size. There are two types of individual features: "single surface" (non-size) and "opposed surface," (size feature).

A "single surface" individual feature, for example, is the flat surface extremity of a size dimension, or a surface given a "flatness" tolerance, etc.

An "opposed surface" individual feature of size, for example, is the cylindrical feature shown under the RULE 1 explanation (pages 22–24), or a specified "circularity" or "cylindricity" tolerance, etc. Specified "parallelism," which involves opposed surfaces, becomes a variety of interrelationship but is considered to retain the connotation of an individual feature and is subject to the confinement of form error within the perfect form at MMC boundary.

Interrelationship (Related) Feature. An "interrelationship" feature involves a relationship of two or more individual features and always requires specified geometric controls and a datum or datum reference frame. "Perpendicularity" is the most predominant example of an interrelationship type feature. One exception to the need for a datum would be the use of "straightness" RFS or MMC applied on a ϕ basis to the part diameter size dimension. In this instance, an application of straightness (all longitudinal elements assumed collectively to establish a *size* feature consideration) would change from the category of "individual" feature to that of "interrelationship" feature because of the interrelationship of size and form which results in a "virtual condition" size exceeding the MMC perfect form boundary.

Due to the variety of applications in which a "feature" can be used, we cannot clearly define it without relating it to a specific drawing requirement. Just as it would be difficult to describe a form or location tolerance application without a drawing, we need pictorial representation and specification of the involved requirements to clearly "see" the "feature."

A drawing illustrates the "feature" relationship within the specification and clarifies the functional intent of the requirement.

See the drawing that follows as an example.

EXAMPLE

Question   How many features are on this part?

Answer   At least thirteen. However, they need not and cannot be identified until the drawing specification is completed.

Question   What are the features on the part shown below in its completed version?

FEATURES IN THIS RELATIONSHIP ARE DATUM FEATURE (PLANE), "D", AND DATUM FEATURES (WIDTHS) "E" & "F" AND THE POS'N TOLERANCE HOLE. NOTE THAT DATUM FEATURE "A" SURFACE IS ONLY ONE-HALF OF DATUM FEATURE "E"; THE SAME SURFACE IS USED IN TWO SEPARATE WAYS IN ESTABLISHING FEATURES.

FEATURES IN THIS RELATIONSHIP ARE DATUM FEATURE (PLANE) "A" AND THE SURFACE HELD IN PAR. WITH IT.

FEATURES IN THIS RELATIONSHIP ARE DATUM FEATURES (PLANES) "B", "C", "D" AND THE 2 HOLES POSITIONALLY TOLERANCED.

POSITION TOLERANCE CONVERSION TO TOOL TOLERANCE

The question is often raised concerning relationship between a part dimensioned and toleranced with positional tolerance and the tooling necessary to produce the part. Positional tolerancing of cylindrical features normally derives cylindrical tolerance zones, whereas the tooling for the part can be based upon X and Y movement in its design or construction. The two geometrically opposed concepts are at times misunderstood as being incompatible with one another.

This point is often also used as an argument of resistance against the use of positional tolerancing using cylindrical tolerance zones. The uninformed fail to make the important distinction between the necessity (and normal practice) of defining the part *end-product* (reflecting its function) on the engineering drawing as the objective, rather than *methods* of manufacture of the part or its tooling.

Since, however, most hard tooling is produced in a shop provided with machinery based upon conventional X and Y axis movement, the methods of translating requirements between the product drawing and the tooling mode is important.

In the example on the following page, a sample part is shown using position tolerancing of three holes. Below the part representation is a typical box type drill jig design, which for purposes of our explanation, is the method used in producing the holes in this part during its manufacture.

Note that the X and Y nominal coordinates are readily transferable to the jig design from the product design. The nest surface and rails in the jig provide the datum location and origination of the jig coordinate dimensions to the drill bushing holes. The part clamping device in the jig must hold the part down against its primary datum A (for orientation), against the datum B rail (for orientation and location), and in contact with the third datum (for location).

The matter of translating (converting) the product positional tolerance (ϕ .007) to an X and Y value is illustrated in the jig drawing.

Imagine that in the evolutionary process from part drawing to part build, a tool order is written from the processing engineer to the tool engineer. This tool order requires the tool engineer to design a drill jig capable of producing 5,000 parts. Suppose as a hypothetical situation, that this tool order specifies, within a company standard or a contractual requirement, an "X" class tool to set the life, quality, and thus the percentage of product tolerance which can be used for this part.

The tool engineer simply converts as shown the total product design positional tolerance (ϕ .007) to the equivalent total coordinate ± tolerance (.005) based upon standard conversion factors (.7 X .007) and derives .005. Reducing the .005 to 20% usable total tolerance for the "X" class jig derives .001 or ±.0005 coordinate tooling tolerance for each dimension in X and Y.

The lower illustration recaps the dimension and tolerance allocation in one plane for clarification.

This reasoning of conversion from positional tolerancing on the product drawing to equivalent coordinate ± tolerancing may be applied to other situations in the process from design to building of parts, e.g. building prototype parts without hard tooling, building temporary tooling without tool design, and inspecting prototype parts.

Note that the transition from positional tolerance to coordinate tolerance is simply done and illustrates the flexibility inherent in the use of this system.

POSITION TOLERANCE CONVERSION TO TOOL TOLERANCE

AS DRAWN

PART CLAMPED
AGAINST DATUM
SURFACES

DRILL BUSHINGS (3)
FOR .201 DRILL

∅.007 POSN. TOL

∅.007 X .7 = .0049 .005

TOOL TOL CLASS X =20%
PROD TOL
20% OF .005 =.001
= ±.0005

XIII Permitted use of ISO Limits and Fits System methods

USE OF ISO LIMITS AND FITS SYSTEM

Limits and fits may be applied per ISO practices; see ANSI B4.2 for complete detailed information on this system.

Example of use:

See Pages 178, 179 for calculation of positional tolerances. (Also ANSI Y14.5M–1982, Appendix B)

Depending upon preference, the optional methods of indicating the limits and fits sizes presented below may be substituted in the above examples:

Part 1	20.000 20.052	(20 H9)	10.036 10.000	(10 H9)
Part 2	19.935 19.883	(20 d9)	9.960 9.924	(10 d9)
Part 1	20 H9		10 H9	
Part 2	20 d9		10 d9	

XIV Symbolic methods of indicating common word requirements are optional:

WORDS	SYMBOL	EXAMPLE	PAGE REFERENCE GM 11
C Bore or Spotface	⊔		Page 202 (lower)
Depth or deep	⊽	4X ⌀.250 ±.005 ⊔⌀.375 ±.020 ⊽ .190 ±.005	Pages 201, 202, 203
Quantity	X		Entire text
C Sink	∨	3X ⌀.250 ±.005 ∨⌀.375 ±.010	None — For Information
Square	□	3X □	Pages 210, 211
All Around	(Symbol)		Page 72
Conical Taper		.010:1	None — For Information
Flat Taper Slope		.020 ±.020:1	None — For Information
Reference Ref }	( )	(.900)	None — For Information
Radius	R	R.300 ±.003	Page 75
Spherical Radius	SR	SR.300 ±.003	None — For Information
Spherical Diameter Spherical Dia	S⌀	S⌀.300 ±.003	None — For Information
Arc Length		2.100	None — For Information

INDEX